Lecture Notes in Mathematics

Edited by A. Dold and B. Eckmann

647

Louis J. Ratliff, Jr.

Chain Conjectures in Ring Theory

An Exposition of Conjectures on Catenary Chains

Springer-Verlag
Berlin Heidelberg New York 1978

Author
Louis J. Ratliff, Jr.
Department of Mathematics
University of California
Riverside, CA 92521
U.S.A.

AMS Subject Classifications (1970): 13 A 15, 13 B 20, 13 C 15, 13 H 99
Secondary: 13 B 25, 13 J 15

ISBN 3-540-08758-3 Springer-Verlag Berlin Heidelberg New York
ISBN 0-387-08758-3 Springer-Verlag New York Heidelberg Berlin

Printing and binding: Beltz Offsetdruck, Hemsbach/Bergstr.
2141/3140-543210

To my Mother and Step-Father

Ruth and Earl McCracken

ACKNOWLEDGMENTS

I want to thank Steve McAdam and David E. Rush for many helpful and stimulating conversations concerning the chain conjectures. Their help shows up in many (frequently indiscernable) ways in these notes.

I want to express my sincere gratitude to the National Science Foundation for their support of my research on these problems (Grants 28939, 28939-1, 28939-2, MPS71-02929-A03, MCS76-06009, and MCS77-00951, and to the University of California, Riverside for my Sabbatical year in residence (1976-77) in which I essentially completed writing these notes.

Finally, I am very grateful to Mrs. Jane Scully for her very excellent typing and her constant good humor and patience.

PREFACE

These notes are concerned with a number of open problems, the catenary chain conjectures, some of which are of quite long-standing. In the hope of attracting some new people to do research on these problems, an attempt has been made to make these notes understandable to non-experts in this area. This shows up partly in the history given for the terminology (Chapter 1), the discussion of the recently solved problems (Chapter 2), the explanation of where the conjectures come from and their history (Chapters 4 - 13), the examples (Chapter 14), and the discussion of the related open problems (Chapter 15). Also, numerous references have been included in the proofs, and many of the referenced results are summarized in Appendix A to help decrease the amount of time the reader must spend searching for a result in research journals. For those more familiar with this area, there are a number of new conjectures given, many implications between the conjectures are shown to hold, numerous equivalences of the more important conjectures are proved, and an extensive bibliography is included. The chain conjectures are interesting and important problems, and it is my hope that the material in these notes will help in making these conjectures more widely known and in determining which of the conjectures are true.

TABLE OF CONTENTS

CHAPTER 0
INTRODUCTION

All rings in these notes are assumed to be commutative with an
identity. The definitions concerning the various chain conditions on
a ring are given in (1.1), and the undefined terminology is, in general,
the same as that in M. Nagata's Local Rings [N-6]. (We also use ⊂
as in [N-6] to mean proper subset.)

Many conjectures concerning saturated chains of prime ideals in
Noetherian domains and in integral extension domains of such a ring
have appeared in the literature. These, and some similar new conjectures, are collected together in these notes in order to see what implications exist between them and to give a number of equivalences of the
more important ones. Besides this, there is given: a brief history
of the terminology and a brief list of some of the most important results that have recently been proved in this area; a summary of where
the conjectures have previously appeared in the literature together
with the known results concerning them; some examples to show that the
conjectures hold for certain classes of local domains and, on the other
hand, to show that there are quasi-local domains for which the conjectures do not hold; a brief list of some related open problems; and, an
extensive bibliography of papers in this area.

The chain conjectures say that one or another of the various chain
conditions holds for certain very large classes of local domains. Now
these same chain conditions are hypothesized to hold in many research
problems in algebraic geometry and commutative algebra, so it is important to determine whether or not these conjectures hold, and any results to this effect should be of quite general interest.

As already noted, the main purpose of these notes is to establish
implications between these conjectures and to prove some equivalences
of the more important of the conjectures. There are a number of reasons

why consideration of these implications and equivalences are of interest and importance, but only two will be mentioned. The first is that if (and when) one or another of these conjectures is shown to hold, then the conjectures which are implied by it, together with the statements equivalent to it, will also be known to hold and will be readily available for use. The second (and more important) reason is that the known conjectures have, to date, withstood the many efforts to settle them, so it is hoped that possibly one of the new conjectures (or one of the new equivalences) will be more readily decidable. (This is a real possibility, since some of the equivalences to the conjectures sound so reasonable that they clearly should hold, and since the equivalences vary over a fairly wide range of concepts, such as: integral extension domains; valuation rings; Rees rings; associated graded rings, polynomial rings; Henselizations; completions; analytically independent elements; and, ideals of the principal class.) And if this turns out to be the case, then some new insight into the other conjectures should result and be of help in future work in this area.

Chapter I contains the definitions of the various chain conditions for a ring, some comments on the history of the terminology is given, and some of the basic facts concerning these concepts are listed.

In Chapter 2, a brief summary of some of the recently proved important results in this area is given.

Chapter 3 contains the conjectures (both old and new), and in this chapter a number of implications between these conjectures are established. It turns out that almost all the conjectures lie between (implicationwise) the Chain Conjecture (3.3.2) and the Catenary Chain Conjecture (3.3.8). (They all lie between (3.3.1) and the Normal Chain Conjecture (3.3.9).) Then, to indicate how things currently stand, two additional statements are also considered. One is somewhat stronger than (3.3.1) and is false, and the other is somewhat weaker than the

Normal Chain Conjecture and is true.

Chapters 4 - 12 have the same general form. In each of these chapters one of the more important conjectures from Chapter 3 is singled out for attention, and there is given some indication of where, when, and why it arose. Then a summary of where it has appeared in the literature and of the results that have previously been proved concerning it is also given, and, finally, a number of new equivalences of it are proved.

Chapter 13 is concerned with (3.3.1) - the one conjecture in Chapter 3 that I think may not be true - and a brief explanation of why I have doubts about this conjecture is given.

Chapter 14 contains some examples to show that the conjectures hold for certain classes of local domains, but that they do not hold for all quasi-local domains, and in Chapter 15 some related open problems are briefly discussed.

In Appendix A quite a few characterizations of certain of the chain conditions $(H_i$-local ring, first chain condition, etc.) are listed and references to where the characterizations were proved are also given. This appendix is a very important part of these notes for at least three reasons. The first is that throughout these notes it is necessary to make reference to many known results in the literature concerning the various chain conditions, and since many of these are explicitly stated in this appendix (with an appropriate reference), we usually give a reference to the result in this appendix rather than to the research paper in which the result is proved. (Hopefully, this will diminish the amount of time the reader must spend in searching research papers to find a referenced result.) A second (and more important) reason is that a number of the equivalences of one or more of the various chain conjectures in Chapters 4 - 13 are in terms of these chain conditions, and so each of these characterizations gives rise to a different (in

appearance) equivalence of the chain conjecture under consideration.
Finally, these lists can also be used to give a number of equivalences
of the other conjectures mentioned in Chapter 3 that are not considered
in detail in Chapters 4 - 13.

Appendix B is concerned with M. Nagata's example(s) to show that
the answer to the chain problem of prime ideals (see the introduction
to Chapter 2) is no. A brief description of the examples is given,
some of their properties are established, and some additional uses
(that have previously appeared in the literature) of these examples is
also given.

The Bibliography is in two parts. The first part lists the sev-
enty books and research papers which are referred to in these notes,
and then follows a Supplemental Bibliography that lists additional important papers in this area which were not specifically referenced in
these notes. (The papers [1,2,3,4,5,7,19,20,22,36,43] in the Supplemental Bibliography are definitely non-Noetherian in flavor and are
mainly concerned with a different aspect of the dimension theory of
rings. However, a number of the methods and results in these papers
are of use for the chain conjectures, so it was felt that these papers
should be included in the Supplemental Bibliography.)

As noted in the Preface, an attempt has been made to make these
notes understandable to non-experts in this area, and I hope the material in them will help attract new people to do research in this area.

DEFINITIONS AND BASIC RESULTS

In this chapter, the various chain conditions on a ring are defined and some comments on the history of the terminology are given. Following this, a few basic results on these definitions that are frequently used in this paper are listed in (1.2) - (1.5).

(1.1) contains the definitions that will be needed in what follows. At least a few of these definitions will probably be new to the reader, since some of the concepts are quite new.

(1.1) DEFINITION. Let A be a ring, and let a = altitude A .

(1.1.1) A chain of prime ideals $P_o \subset P_1 \subset \cdots \subset P_k$ in A is a saturated chain of prime ideals in A in case, for each $i = 1,\ldots,k$, height $P_i/P_{i-1} = 1$. The length of the chain is k . The chain is a maximal chain of prime ideals in A in case it is a saturated chain of prime ideals such that P_o is minimal and P_k is maximal. In this case it will be said that $P_o \subset \cdots \subset P_k$ is a mcpil k in A (that is, a maximal chain of prime ideals of length k) .

(1.1.2) A satisfies the first chain condition for prime ideals (f.c.c.) in case each maximal chain of prime ideals in A has length = a .

(1.1.3) A is catenary in case, for each pair of prime ideals $P \subset Q$ in A , $(A/P)_{Q/P}$ satisfies the f.c.c.

(1.1.4) A satisfies the second chain condition for prime ideals (s.c.c.) in case, for each minimal prime ideal z in A , each integral extension domain of A/z satisfies the f.c.c. and depth z = a .

(1.1.5) A satisfies the chain condition for prime ideals (c.c.) in case, for each pair of prime ideals $P \subset Q$ in A , $(A/P)_{Q/P}$ satisfies the s.c.c.

(1.1.6) A is level in case all maximal ideals in A have the

same height and all minimal prime ideals in A have the same depth.
A is taut (respectively, taut-level) in case, for each prime ideal P
in A , height P + depth P $\in$ {1,a} (respectively, = a) .

    (1.1.7) A satisfies the $1\frac{1}{2}$ chain condition for prime ideals
(o.h.c.c.) in case A is taut and, for each minimal prime ideal z
in A , the integral closure of A/z satisfies the c.c.

    (1.1.8) A is an H_i-ring (i $\geq$ 0) in case, for each height i
prime ideal P in A , depth P = a - i (that is, height P +
depth P = a) . (For i = 1 , it will usually be said that A is an
H-ring, instead of saying that A is an H_1-ring.)

    (1.1.9) A is a C_i-ring (i $\geq$ 0) in case A is an H_i-ring,
an H_{i+1}-ring, and, for each height i P $\in$ Spec A , all maximal ideals
in the integral closure of A/P have the same height (= depth P = a
- i) .

    (1.1.10) A is a GB-ring in case adjacent prime ideals in each
integral extension ring of A contract in A to adjacent prime ideals.
(P $\subset$ Q are adjacent in case height Q/P = 1 .)

    (1.1.11) If A is an integral domain, then A satisfies
the altitude formula (respectively, the dominating altitude formula) in case, for all finitely generated integral domains B over A
and for all P $\in$ Spec B (respectively, for all P $\in$ Spec B such that
P$\cap$A is a maximal ideal), altitude B_P + trd (B/P)/(A/(P$\cap$A)) = altitude
$A_{P\cap A}$ + trd B/A , where trd D/C denotes the transcendence degree of the
quotient field of the integral domain D over the quotient field of its
subdomain C .

    (1.1.12) If A is a semi-local (Noetherian) ring, then A is un-
mixed (respectively, quasi-unmixed) in case all (respectively, all minimal) prime divisors of zero in the completion of A have the same depth.

    (1.1.13) If A is a local ring with maximal ideal M , then a
special extension $(A^S$; M_1 , M_2) of A is a principal (or, simple)

integral extension ring $A^S = A[x]$ of A that has exactly two maximal ideals $M_1 = (M,x)A^S$ and $M_2 - (M,x-1)A^S$, and $M_1 \cap M_2 = M$.

(1.1.14) A' denotes the integral closure of A in its total quotient ring, and if A is semi-local, then A^H and A^* denote the Henselization and completion of A , respectively.

The "mcpil k" terminology was introduced by S. McAdam and myself in 1976, in [RMc], since the phrase for which it is an abbreviation is quite long and occurred very frequently in [RMc]. (Of course, maximal chains of prime ideals had been considered by many authors in many papers prior to [RMc].) Some known equivalences of the existence of a mcpil k in a local ring are listed in (A.5), and some additional related results are given in (1.5).

The f.c.c., s.c.c., c.c., quasi-unmixed, unmixed, and altitude formula are conditions that are well known to hold for local domains of classical algebraic geometry. The first papers I know of in which this terminology is used and in which these concepts are studied in their own right are M. Nagata's 1956 paper [N-3], for: f.c.c., s.c.c., quasi-unmixed, and unmixed; and his 1962 book [N-6], for: c.c. and altitude formula. (He used "dimension formula" for "altitude formula" in 1959, in [N-5], and this is also used in [ZS-2]. This formula generalizes the classical result that if P is a prime ideal in $F_n = F[X_1,\ldots,X_n]$, where F is a field, then height P + depth $P = n$ and depth $P = $ trd $(F_n/P)/F$.) The French school terminology for these concepts is: "bi-equidimensional" [G-1, (16.1.4)], for "f.c.c.;" "formally equidimensional" [G-2, (7.1.9)], for "quasi-unmixed;" "strictly formally equidimensional" [G-2, (7.1.9) and (7.2.1)], for "unmixed;" and, "universally catenary" [G-2, (5.6.1) and (5.6.2)], for "altitude formula" (but universally catenary is defined for Noetherian rings while altitude formula is defined for integral domains, so the matching of the terminology is not perfect in this case). (See (2.6).) There

does not seem to be very good corresponding terminology for "s.c.c."
or for "c.c."

The "catenary" terminology is also due to the French school - for
example, see [G-1, (16.1.4)]. (Nagata did not introduce terminology
for this condition in [N-3] or in [N-6].) This condition on a ring is
called the "chain condition for prime ideals" in [ZS-2, p. 326], and
I used "saturated chain condition for prime ideals" in [R-2] and [R-3].
(The French word caténaire translates to catenary, chainlike, or chain,
so this is why I used the parentheses in the title of these notes.)

A few basic results on the f.c.c., s.c.c., catenary, and c.c. conditions are given in (1.2) and (1.3), and quite a few known characterizations of a local domain that satisfies the f.c.c. or the s.c.c. are
listed in (A.9) and (A.11), respectively.

I introduced the "dominating altitude formula" terminology in
1972, in [R-5], in order to show that two different definitions in the
literature for a Noetherian domain to satisfy the altitude formula are
equivalent.

The word "level" has been used to mean various things in the literature. For example, D. Rees used it to mean quasi-unmixed, in [Re],
and it has elsewhere been used to mean that all minimal prime ideals
in a ring have the same depth, and, on the other hand, that all maximal
ideals in a ring have the same height.

The "taut" and "taut-level" conditions were introduced by S. McAdam
and myself in [McR-2]. However, the concept of such a ring can clearly
be traced back to I. S. Cohen's 1954 paper [C-2]. (Concerning this,
see (2.2).) The reason for considering these two conditions is that
an integral extension domain of a catenary local domain is taut, by
[McR-2, Proposition 12], and it was hoped that by studying this condition it could be shown that a taut (respectively, taut-level) semi-
local domain is catenary (respectively, satisfies the f.c.c.). (Con-

cerning this, see Chapter 10.)   A number of known characterizations of
a taut local ring and of a taut semi local domain are listed in (A-7)
and (A.8), respectively.

The "o.h.c.c." was introduced by M. E. Pettit, Jr. and myself in
[RP], and therein it is shown that this condition on a local ring stands
in relation to the s.c.c. much as the taut condition stands in relation
to the f.c.c. A deeper study of this condition is made in [R-18]. (The
terminology for this condition is not particularly good, but the condition does fall intermediate to the f.c.c. and the s.c.c. for a local domain, by [R-18,(2.3)]. A number of known characterizations of a semi-
local domain that satisfies the o.h.c.c. are listed in (A.10), and two
basic results on this condition are given in (1.4).

I introduced the "H" condition for a ring in 1971, in [R-4, Section 4], and it was noted in [R-6] that the reason H was used is it
was hoped that every Henselian local domain is an H-domain. (If this
is true, then the Chain Conjecture (3.3.2) holds - see (4.1.1) ⇔
(4.1.4).)

M. E. Pettit, Jr. generalized the "H" condition to the "H_i" and
the "C_i" conditions in 1973, in [P]. These conditions are finer conditions on a local domain than the s.c.c., and so it was hoped that a
study of these finer conditions would yield some new information on
chains of prime ideals that would help in the study of the chain conjectures. A number of known characterizations of a local ring that
satisfies one of these conditions are listed in (A.3) and (A.4).

I introduced the "GB" terminology in [R-16], but the concept of a
GB-ring can be traced back to W. Krull's 1937 paper [Kr]. (Concerning
this, see (2.1).) It turns out that a local domain satisfies the s.c.c.
if, and only if, it is catenary and is a GB-ring, by [R-16, (3.10)].
Now the catenary condition has been deeply studied in a number of research papers, but only widely scattered results on the GB condition

were known prior to [R-16] and [R-19]. (The GB stands for "going be-tween" - it readily follows from the definition that a ring A is a GB-ring if, and only if, whenever $P \subset Q$ are prime ideals in an inte-gral extension ring B of A such that there exists a prime ideal p^* in A such that $P \cap A \subset p^* \subset Q \cap A$, then there exists a prime ideal P^* in B such that $P \subset P^* \subset Q$. This condition on a ring is some-what comparable to the GD and the GU conditions.) An open question concerning a GB-local domain is mentioned in (15.3), and a list of known characterizations of GB-rings is given in (A.6).

Finally, I introduced the "special extensions" terminology in 1973, in [R-6, Section 4]. M. Nagata's examples [N-6, Example 2, pp. 203-205] (see Appendix B) involve special extensions, and I used this concept to give a characterization of the Catenary Chain Conjecture, (3.3.8), in [R-6, (4.7)]. It will be seen in these notes that a number of the chain conjectures can be characterized in terms of special extensions, and a considerably deeper study of such extension rings is given in [D].

In (1.2) - (1.5) some needed basic facts on the concepts in (1.1) will be given. Before this, however, to help those for whom the ter-minology is new understand some of the relationships between several of these definitions, we make a few comments on them.

It is clear that s.c.c. $\Rightarrow$ f.c.c. $\Rightarrow$ taut-level $\Leftrightarrow H_i$ for all $i \geqq 0$, that $C_i \Rightarrow H_i$, and that a ring A is a C_i-ring and an H_i-ring for all $i \geqq$ altitude A .

A taut-level ring A is a taut ring, and a taut ring is either taut-level or fails to be taut-level only because there exists a height one maximal ideal in A and/or there exists a depth one minimal prime ideal in A . Also, a taut local ring is an H_i-ring, for $i \geqq 1$ (and is an H_0-ring if, and only if, it is taut-level), and a taut semi-local ring is an H_i-ring, for $i \geqq 2$.

It readily follows from [R-5, Theorem 2.2] (see (A.9.1) $\Leftrightarrow$ (A.9.5))

that the following statements are equivalent for a local domain R :
R satisfies the f.c.c.; R is catenary; R is taut-level; R is
taut; and, R is an H_i-ring, for i = 1,...,a-2 , where a = altitude R .

Also, by [RP, (3.11)], the following statements are equivalent
for a local domain R : R satisfies the s.c.c.; R satisfies the c.c.;
R satisfies the o.h.c.c. and R' is level; and, R is a C_i-ring,
for i = 0,1,...,a-2 , where a = altitude R .

By [N-6, Example 2, pp. 203-205] in the case m = 0 , there exists
a local domain R that satisfies the o.h.c.c., but not the s.c.c.
(See (B.3.2).) Also, for m ≧ 0 , there are local domains that are
H_i-rings (respectively, C_i-rings) for i = m+1,...,a = altitude R ,
but are not H_j-rings, for j = 1,...,m (respectively, C_j-rings, for
j = 0,1,...,m .) (See (B.4.1) and (B.4.2).)

We now give four remarks that contain basic facts on the concepts
in (1.1) that are needed in the remainder of these notes. All the
statements in the first two of these follow quite readily from the
definitions. (A sketch of the proofs of these statements can be found
in [R-4, Remarks 2.22 - 2.24].) (1.2) is concerned with catenary rings
and the f.c.c.

(1.2) REMARK. The following statements hold for an integral domain A :

(1.2.1) If A is catenary, then A_S/IA_S is catenary, for all
ideals I in A and for all multiplicatively closed subsets S (0 ∉
S) in A .

(1.2.2) A is catenary if, and only if, A_M is catenary, for all
maximal ideals M in A . (Note, if height M < ∞ , then A_M is catenary if, and only if, A_M satisfies the f.c.c.)

(1.2.3) If altitude A < ∞ , then A satisfies the f.c.c. if,
and only if, A is level and A_M is catenary, for each maximal ideal

M in A , and this holds if, and only if, A is level and catenary.

(1.2.4) If B is an integral extension domain of A and B
satisfies the f.c.c., then A satisfies the f.c.c.

(1.3) is concerned with the c.c. and the s.c.c. It shows that
c.c. stands in relation to s.c.c. much as catenary stands in relation
to f.c.c.

(1.3) REMARK. The following statements hold for an integral domain A :

(1.3.1) If A satisfies the c.c., then A_S/IA_S satisfies the
c.c., for all ideals I in A and for all multiplicatively closed
subsets S (0 ∉ S) in A .

(1.3.2) A satisfies the c.c. if, and only if, A_M satisfies
the c.c., for all maximal ideals M in A . (Note, if height M < ∞ ,
then A_M satisfies the c.c. if, and only if, A_M satisfies the s.c.c.)

(1.3.3) If altitude A < ∞ , then A satisfies the s.c.c. if,
and only if, A is level and A_M satisfies the c.c., for each maximal
ideal M in A , and this holds if, and only if, A is level and
satisfies the c.c.

(1.3.4) If B is an integral extension domain of A and B
satisfies the s.c.c., then A satisfies the s.c.c.

(1.3.5) If A satisfies the s.c.c. (respectively, the c.c.),
then A satisfies the f.c.c. (respectively, A is catenary).

Because of (1.2.4) and (1.3.4), the reader might expect that if
A ⊆ B are integral domains such that B is integral over A and B
is catenary (respectively, satisfies the c.c.), then A is catenary
(respectively, satisfies the c.c.). For the c.c., this is, in fact,
stated in [N-6, (34.2)]. However, neither of these statements is true
(even when A is a local domain and B is a finite A-algebra), as
follows from [N-6, Example 2, pp. 203-205]. (See (B.3.4) and (B.5.1).)

On the other hand, if B is catenary (respectively, satisfies the c.c., and if, for each maximal ideal M in B , height M∩A ▪ height M < ∞ , then Λ is catenary (respectively, satisfies the c.c.), as follows from the statements in (1.2) (respectively, (1.3)).

(1.4) gives two facts concerning the o.h.c.c. (It is shown in (B.3.12) that the restriction to semi-local rings is necessary in (1.4).)

(1.4) REMARK. Let R be a semi-local (Noetherian) ring that satisfies the o.h.c.c. Then the following statements hold:

(1.4.1) [R-18, (2.3)]. R is catenary.

(1.4.2) [R-18, (2,9) and (2.10)]. For each $P \in$ Spec R, R_P and R/P satisfy the o.h.c.c.

(1.5) is concerned with the existence of a mcpil n in an integral extension domain.

(1.5) REMARK. [RMc, (2.14)]. The following statements are equivalent for a local domain (R,M) :

(1.5.1) There exists a mcpil n in some integral extension domain of R .

(1.5.2) There exists a minimal prime ideal z in the completion R^* of R such that depth z = n .

(1.5.3) There exists a mcpil n + k - depth Q in R_{kQ} , where $R_k = R[X_1, \ldots, X_k]$, $Q \in$ Spec R_k , and $MR_k \subset Q$.

(1.5.4) There exists a mcpil n+1 in $R[X]_{(M,X)}$.

(There are two additional statements equivalent to the statements in (1.5) that are given in [RMc, (2.14)], but they will not be needed here.)

CHAPTER 2

SOME RECENTLY SOLVED PROBLEMS

In any discussion of the various chain conditions for prime ideals
in a Noetherian ring, there are two results which stand above all other
such results. The first is due to I. S. Cohen in 1946, in [C-1, Theorem 19], and it shows that a complete local ring is catenary (1.1.3).
Therefore, since a local domain is a dense subspace of its completion,
a natural question to ask is if every local domain is catenary. This
is the chain problem of prime ideals, and its solution by M. Nagata,
in 1956, is the second result just alluded to. He showed, in [N-3,
Theorem 1 and Section 3], that the answer is yes for quasi-unmixed local domains (for example, regular local rings or homomorphic images of
Macaulay rings), but, in general, the answer is no. To show that the
answer is no, he constructed a family of local domains R such that
altitude $R = r + m + 1$ ($r \geqq 1$ and $m \geqq 1$) such that the short chains
in R all have length $m + 1$ and are due to a maximal ideal M' in
the integral closure R' of R such that height $M' = m + 1$. (His
examples included the case $m = 0$, and for this case R is catenary
and there exists a height one maximal ideal in R'.) To date, his
examples are (essentially) the only known counter-examples to the chain
problem. (The examples are reproduced in [N-6, Example 2, pp. 203-205],
and similar examples are given in [G-2, (5.6.11)], [M, (14.E)], and
[ZS-2, Example, pp. 327-329]. Appendix B contains a brief description
of these examples together with some facts concerning them.)

Since 1956 (and because the answer to the chain problem is no), a
number of other problems and conjectures concerning chains of prime
ideals in Noetherian domains have appeared in the literature, and some
of these have recently been settled. A few of the more important suc-
cesses will now be mentioned, and we begin by noting a non-Noetherian
solution of a problem of considerably longer standing.

(2.1)   In 1937, in [Kr, p. 755], W. Krull asked the following
question: if B is an integral domain that is integral over its integrally closed subdomain A , and if P ⊂ Q are adjacent prime ideals
in B , then does it follow that P∩A ⊂ Q∩A are adjacent prime ideals?
(That is, is every integrally closed domain a GB-ring (1.1.10)?)   In
1972, I. Kaplansky gave a negative answer to this question in [K].   In
[K], the ring A is not the integral closure of a Noetherian domain
(nor is it a Krull domain), and the problem is still open for this case.
(If the Chain Conjecture (3.3.2) holds, then the answer, for this case,
is yes, by (3.6.1) ⇒ (3.6.4).)

(2.2)   In 1956, in [Y], M. Yoshida asked if a taut-level (1.1.6)
local domain R must be catenary.   (This question was clearly suggested
by I. S. Cohen's discussion of the chain problem in 1954, in [C-2, p.
655], and by M. Nishi's 1955 paper [Ni].)   In 1972, in [R-5, Theorem
2.2], I gave an affirmative answer to this.   And, in 1973, in [McR-2,
Proposition 7] and in [R-14, (2.15)], S. McAdam and I showed that if
R is a taut-level local ring, then R satisfies the f.c.c. (1.1.2).
(Related to this, in 1972, in answer to a question asked in [R-5, Remark
2.6(iv)], W. Heinzer gave an example of a taut-level quasi-local domain
that is not catenary.   (See (14.6).)   Some additional results related
to this are given in (2.8).)

(2.3)   In 1956, in [N-3, Problem 1, p. 62], M. Nagata asked (*):
can (0) in the completion R^* of a local domain (R,M) have imbedded
prime divisors?   And, in 1959, in [N-5, Section 4], he asked if the
M-transform of R (= UM^{-n} = [B, Corollary 1.6] $R^{(w)}$ = ∩{R_p ; p ∈ Spec R
and p ≠ M}) must be a finite R-algebra when the integral closure R'
of R is quasi-local, and he then commented that if the answer was yes,
then he could prove that the answer to (*) was no and that the Chain
Conjecture holds, but if the answer was no, then it was almost certain

that an example of such a ring could be used to show the answer to (*)
was yes. (In 1970, in [F-M, p. 120], M. Flexor-Mangeney showed that
(*) and the finiteness of $R^{(w)}$ are, in fact, equivalent problems;
that is, the answer to (*) is no if, and only if, $R^{(w)}$ is always a
finite R-module, when R' is quasi-local.) In 1970, in [FR, Proposition 3.3], D. Ferrand and M. Raynaud gave an example of a local domain
R such that R' is local and $R^{(w)}$ is not a finite R-module, and
the zero ideal in R^* has an imbedded prime divisor. (Of course, for
this R , $R \neq R'$ (since $R^{(w)}$ is not finite over R) , and it is
still an open problem if the answer to (*) is yes, when R is integrally closed.) Related to (*), in 1948, in [Z], O. Zariski asked if
every normal local domain is analytically irreducible. In 1955 and
1958, in [N-2] and [N-4] (whose titles are self-explanatory), M. Nagata
gave a negative answer. Also related to (*), in 1953, in [N-1, Conjecture 1], M. Nagata asked if there exists a Henselian local domain R
such that (0) in R^* is not primary, and he gave an affirmative answer to this in 1958, in [N-4] - and [FR, Proposition 3.3] showed that
(0) in such R^* can even have an imbedded prime divisor. (All of
these questions are related to the chain conjectures, since it is known
[R-2, Theorem 3.1] that a local domain is quasi-unmixed (1.1.12) if,
and only if, it satisfies the s.c.c. (1.1.4). Concerning this, see
(2.5).)

(2.4) In [N-3, Problem 1, p. 62], it was asked if (0) in R^*
can have an imbedded prime divisor when all minimal prime ideals in
R^* have the same depth; that is, does there exist a quasi-unmixed local
domain that is not unmixed (1.1.12)? [FR, Proposition 3.3] (together
with [R-2, Proposition 3.5]) gives an affirmative answer to this.
(Prior to this answer, I proved in 1970, in [R-3, Proposition 5.13],
that the answer is yes if, and only if, there exists a Henselian local
domain R such that $R^{(1)} = \cap\{R_p ; p \in \text{Spec } R$ and height $p = 1\}$ is

not a finite R-algebra. Compare this with the result in [F-M, p. 120]
mentioned in (2.3).) There is a related problem [N-3, Problem 2, p. 62]
which is still open: is a factor domain of an unmixed local domain R
unmixed? A partial answer to this was given by M. Brodmann in 1974, in
[Bro, (5.9)], where it was shown that R/p is unmixed, for all but at
most finitely many height one p ∈ Spec R . This also follows from
[R-7, Remark 3.2(2)] together with [N-6, (18.11)].

(2.5) In 1969, in [R-2, Theorem 3.1], I proved the following conditions are equivalent for a local domain R : R satisfies the altitude formula (1.1.11); R satisfies the s.c.c.; and, R is quasi-
unmixed. And, in 1972, in [R-5, Theorem 3.3], I showed these are also
equivalent to: R satisfies the dominating altitude formula (1.1.11);
thus showing that two different definitions of the altitude formula
which had appeared in the literature are equivalent for Noetherian domains. The question of the equivalence of the first two of these conditions arose in 1959, in [N-5], when M. Nagata gave a proof showing that
if R satisfies the s.c.c., then every locality over R satisfies the
s.c.c. and the altitude formula. Unfortunately, the proof used: R
satisfies the s.c.c. if the integral closure of R satisfies the f.c.c.;
and this is still an open problem - see (3.3.9). Concerning the equivalence of the second and third conditions, in [N-3, Theorem 1] and in
[N-6, (34.6)] it was proved that R satisfies the s.c.c., if R is
quasi-unmixed. But the proof in [N-3] was based on a result that is
closely related to the just noted open problem, and the proof in [N-6]
was not completely clear, so some added details for this implication
were given in [R-1, Corollaries 2.6 and 2.7] and in [R-1, pp. 283-284].

(2.6) In 1965, in [G-2, (7.1.12)(i)], A. Grothendieck asked if
a local ring R which is <u>universally catenary</u> (that is, every finitely
generated R-algebra is catenary) is <u>formally catenary</u> (that is, for

each minimal prime ideal z in R , R/z is quasi-unmixed).   And,
in [G-2, (7.2.10)(ii)], he asked if, in parts (c), (d), and (e) of the
defining theorem of a <u>strictly formally catenary</u> local ring R (that
is, R is universally catenary and $(R/p)^{(1)}$ is a finite R/p-algebra,
for all p ∈ Spec R), the condition: R is universally catenary; could
be replaced by: R is catenary. An affirmative answer was given to
each question in 1970, in Theorems 2.6 and 5.4 in [R-3].

(2.7)   Circa 1970, I. Kaplansky asked if each local domain which
satisfies the s.c.c. is a homomorphic image of a Macaulay ring.   A
negative answer is given by [FR, Proposition 3.3] together with [N-6,
(34.10)].   That is, [FR, Proposition 3.3] gives an example of a quasi-
unmixed local domain R that is not unmixed (see (2.4) above), and
[N-6, (34.10)] says that a factor domain of a Macaulay local ring is
unmixed.   Finally, a quasi-unmixed local domain satisfies the s.c.c.,
by (A.11.1) ↔ (A.11.4).

(2.8)   In 1975, in [Fu-1, Lemma 6] (together with [R-2, Theorem
3.6]), K. Fujita showed that there exists a Noetherian Hilbert domain
that satisfies the f.c.c. but not the s.c.c.   (M. Nagata's examples
[N-3] showed this could happen for a local domain, and Fujita's example
was based on Nagata's. See (B.3.11).) This answers a question I asked
in 1973, in [R-11, (2.19) and (2.20)]. Prior to this I showed, in
[R-11, (2.22)], that there exists a Noetherian Hilbert domain that does
not satisfy the c.c.   (See (B.5.7).) Fujita also gave, in [Fu-1, Proposition, p. 478], a negative answer to a question I asked in 1971, in
[R-4, Remark 2.25] and in [R-14, (2.14)], with an example of a Noetherian Hilbert domain that is taut-level but does not satisfy the f.c.c.
(Compare this with the fact that a taut-level local domain satisfies
the f.c.c., as noted in (2.2).) Following this, S. McAdam showed in
[Mc-2, Theorem 4] that R⟨X⟩ satisfies the f.c.c. if it is taut-level,

where R is a semi-local domain and $R\langle X\rangle = R[X]_S$, with $S' = R[X] -$
U{N ; N and N∩R are maximal ideals}. (The last two results are related
to the Taut-Level Conjecture (3.9.5).) Fujita also showed in [Fu-1,
Propositions, pp. 482 and 484] that there exists an H-Noetherian domain (1.1.8) that is not taut and there exists an H-quasi-local domain that is not catenary. The first of these results answers part of
the question I asked in 1971, in [R-4, Remark 2.25], and both of these
results are related to the H-Conjecture (3.3.6).

(2.9) It is known [R-10, (3.1)] that if (R,M) and (S,N) are
local domains such that $R \subseteq S$, N∩R = M , and S is integral over R ,
then there exists a mcpil n (1.1.1) in S if, and only if, there
exists a mcpil n in R . From this, a natural question is if a maximal chain of prime ideals in S must contract to a maximal chain of
prime ideals in R [R-10, (3.15)]. In [R-20, (2.10)] (see (B.4.8)),
I showed that there exist such R ⊂ S such that S has a maximal
chain of prime ideals that does not contract to a maximal chain of prime
ideals in R . (It is pointed out in [R-10, (4), p. 87] that if this
can be shown to hold with R Henselian, then the Chain Conjecture
(3.3.2) does not hold.) [R-20] also contains examples showing that a
number of other things concerning chains of prime ideals in integral
extension domains that were shown by Nagata's examples can, in fact,
be shown in local (rather than semi-local) integral extension domains.

CHAPTER 3

SOME (CATENARY) CHAIN CONJECTURES

In this chapter, many statements (conjectures) concerning saturated chains of prime ideals in a Noetherian domain are considered, and
a number of relationships between these statements are proved. (A diagram showing the major implications that are proved in this chapter is
given on p. 44.) It is unknown which (if any) of the statements hold,
but my feeling is all of them, except (3.3.1), should hold. (Concerning (3.3.1), see Chapter 13.) Two additional statements are also given,
in (3.5), one of which is somewhat stronger than any of the conjectures
and is false, and the other is somewhat weaker than any of the conjectures and is true.

The following two lemmas will be needed quite often in what follows. For the applications of (3.1) in these notes, we will usually
take $S = R'$, but occasionally we will take $S = R$.

(3.1) LEMMA. Let R be a semi-local domain such that altitude
$R > 1$ and let S be a ring such that $R \subseteq S \subseteq R'$. Assume there
exists a height one maximal ideal in S and let b in all height one
maximal ideals in S such that $1 - b$ is in all other maximal ideals
in S . Let $B = R[b,1/b]$ and $C = S[1/b]$. Then the following
statements hold:

(3.1.1) B is a semi-local domain, $B \subseteq C \subseteq B' = R'[1/b]$, and C
has no height one maximal ideals.

(3.1.2) B is a local domain, if R is a local domain.

(3.1.3) A prime ideal P in R[b] (respectively, in S) is lost
in B (respectively, in C) if, and only if, P is a height one maximal ideal.

(3.1.4) B is an H-domain, if R is an H-domain, and the converse holds if there are no height one maximal ideals in R .

(3.1.5) B (respectively, C) is taut-level, if, and only if, R (respectively, S) is taut.

(3.1.6) B (respectively, C) is catenary if, and only if, R (respectively, S) is catenary.

(3.1.7) B' satisfies the c.c. if, and only if, R' satisfies the c.c.

This is proved in detail in [D], so we simply note here that it is proved in much the same way as the local domain case was proved in [R-10, Section 4].

The following lemma, which will be needed in a number of results below, generalizes [R-21, (9.2.1)]. That is, [R-21, (9.2.1)] says that if R is an H-semi-local domain such that altitude $R > 1$, then R' is an H-domain if, and only if, there are no height one maximal ideals in R' , and this is true if, and only if, R' is level. However, we need the more general result that is given in (3.2). It should be noted that the hypothesis depth $p \in \{0,a-1\}$ in (3.2) is satisfied if R is an H-domain, is taut (or taut-level), or satisfies the f.c.c. (or the s.c.c.).

(3.2) LEMMA. Let R be a semi-local domain such that altitude $R = a > 1$. Assume that, for each height one $p \in$ Spec R , depth $p \in \{0,a-1\}$. Then, for each height one $p' \in$ Spec R' , depth $p' \in \{0,a-1\}$; and for each maximal ideal M' in R' , height $M' \in \{1,a\}$. Moreover, the following statements are equivalent:

(3.2.1) R and R' are H-domains (1.1.8).

(3.2.2) R' is level (1.1.6).

(3.2.3) There does not exist a height one maximal ideal in R' .

(3.2.4) There does not exist an integral extension domain of R that has a height one maximal ideal.

(3.2.5) R and R' are C_o-domains (1.1.9).

Proof. Assume first that there are no height one maximal ideals
in R' . Then there are no height one maximal ideals in R (by integral dependence), so R is an H-domain (by hypothesis), and thus R'
is level and is an H-domain, by assumption and [R-21, (9.2.1)].
Therefore the statements concerning height M' and depth p' hold,
and (3.2.3) implies (3.2.2) and (3.2.1).

Now, either of (3.2.1) or (3.2.2) implies (3.2.3), since $a > 1$,
and (3.2.1) together with (3.2.2) imply (3.2.5), since R and R' and
H_o-domains.

(3.2.5) $\Rightarrow$ (3.2.4), since if there exists an integral extension
domain of R that has a height one maximal ideal, then there exists
an integral extension domain S of R such that $S \subseteq R'$ and S has
a height one maximal ideal, by [R-4, Lemma 2.9], so, by integral dependence, there exists a height one maximal ideal in R' , and this contradicts (3.2.5) (since $a > 1$) .

Finally, it is clear that (3.2.4) $\Rightarrow$ (3.2.3).

Now assume that there exists a height one maximal ideal in R'
and let b , S = R' , B , and C be as in (3.1). Then R[b] satisfies the conditions on R , by [Mc-1, Theorem 8], so it follows from
(3.1.1) and (3.1.3) that B = R[b,1/b] is an H-semi-local domain and
B' = C = R'[1/b] has no height one maximal ideals. Therefore, by what
has already been proved, R'[1/b] is level and is an H-domain, so the
statements concerning depth p' and height M' hold, by (3.1.3), q.e.d.

We now begin to consider the various chain conjectures. In (3.3),
nine statements concerning chains of prime ideals in a local domain
are listed, and it is shown that each implies the next. I do not know
if any of the reverse implications hold. There are four additional

conjectures that could be given in (3.3), but we prefer to give them later in these notes. The first two are given in (3.4), and they are separated from (3.3) because they are very closely related to (3.3.3) and (3.3.4) (see (3.4)). The third is the Strong Avoidance Conjecture, (3.8.2), and it is shown in (3.8) and (3.14.1) (and (3.13)) that (3.3.2) ⇒ (3.8.2) ⇒ (3.3.3). The reason for not including (3.8.2) in (3.3) is that we prefer to consider the chain conjectures concerning a local domain first, and then consider the conjectures for the non-local case later in this chapter (from (3.6.4) on). The fourth conjecture that is not given in (3.3) is (15.1), and it is clear that (3.3.9) ⇒ (15.1). One reason for not giving this conjecture till then is that all the other conjectures in these notes are intermediate to (3.3.1) and (3.3.9). Another reason is mentioned preceding (15.1).

It turns out that to prove (3.3) (and a number of other results in this chapter) it is sometimes considerably easier to prove that a given statement implies one of the equivalences (proved in Chapters 4 - 13) of the succeeding statement rather than the succeeding statement itself. When this is the case, we will make use of these equivalences, and then use care in Chapters 4 - 13 to keep away from circular proofs.

Six of the statements in (3.3) are named, and five of these have appeared previously in the literature under these names. The other named conjecture, (3.3.4), and the remaining three conjectures are new in these notes.

(3.3) THEOREM. For the following statements, (3.3.i) ⇒ (3.3.i+1), for i = 1,...,8 :

(3.3.1) If R is a local domain such that, for all nonzero P ∈ Spec R , R/P satisfies the s.c.c., then R is catenary.

(3.3.2) The Chain Conjecture holds; that is, the integral closure of a local domain satisfies the c.c.

(3.3.3) The Depth Conjecture holds; that is, if R is a local domain and P ∈ Spec R is such that height P > 1 , then there exists p ∈ Spec R such that p ⊂ P and depth p = depth P + 1 .

(3.3.4) The Weak Depth Conjecture holds; that is, if R is a local domain and P ∈ Spec R is such that height P = h and depth P = 1 , then there exists p ∈ Spec R such that height p = 1 and depth p ≦ h .

(3.3.5) If (R,M) is a local domain, then the Weak Depth Conjecture holds in $R[X]_{(M,X)}$.

(3.3.6) The H-Conjecture holds; that is, an H-local domain is catenary.

(3.3.7) If R is an H-local domain and R^s is a special extension of R (1.1.13), then $R^s_{M_i}$ is an H-domain, for i = 1,2 .

(3.3.8) The Catenary Chain Conjecture holds; that is, the integral closure of a catenary local domain satisfies the c.c.

(3.3.9) The Normal Chain Conjecture holds; that is, if the integral closure of a local domain R satisfies the f.c.c., then R satisfies the s.c.c.

Proof. To prove that (3.3.1) ⇒ (3.3.2), it suffices, by (4.1.1) ⇔ (4.1.2), to prove that if (3.3.1) holds, then a Henselian local domain satisfies the s.c.c. For this, let R be a Henselian local domain. Then it may clearly be assumed that a = altitude R > 1 . Let (0) ≠ P ∈ Spec R , so R/P is Henselian, and so, by induction on a , R/P satisfies the s.c.c. Therefore, by (3.3.1), R is catenary. Hence, since R is Henselian, R satisfies the s.c.c., by (A.11.14) ⇒ (A.11.1).

Assume (3.3.2) holds, let R be a local domain, and let P ∈ Spec R such that height P > 1 . Then there exists P' ∈ Spec R' such that P'∩R = P and height P' = height P . Let c ∈ P' such

that c is not in any maximal ideal in R' that does not contain
P' , and let p' be a prime ideal in R' such that c ⊂ p' ⊂ P' and
height P'/p' = 1 . Now R' is catenary, by hypothesis and (1.3.5),
so R'/p' is catenary, by (1.2.1). Also, P'/p' is contained in all
maximal ideals in R'/p' (by the choice of c) , so depth P' =
depth P'/p' = altitude R'/p' - height P'/p' = depth p' - 1 . Therefore, p = p'∩R ⊂ P and depth p = depth p' = depth P' + 1 = depth
P + 1 , so (3.3.2) ⇒ (3.3.3).

Assume (3.3.3) holds, let R be a local domain, and let P be
a height h and depth one prime ideal in R . If h ≦ 1 , then clearly
(3.3.4) holds, so assume h > 1 . Then, by (3.3.3), there exists a
prime ideal $P_1 ⊂ P$ such that depth P_1 = depth P + 1 = 2 . Then,
clearly, height P_1 ≦ h - 1 . Therefore (3.3.4) follows from a finite
number of repetitions of this.

It is clear that (3.3.4) ⇒ (3.3.5).

Assume (3.3.5) holds, let (R,M) be an H-local domain, and suppose that R is not catenary. Then, by (3.1.1), (3.1.2), (3.1.4),
and (3.1.6) (with S = R') , to prove that (3.3.6) holds, it may be
assumed that there does not exist a height one maximal ideal in R' .
Therefore, by (A.3.1) ⇒ (A.3.17), $D = R[X]_{(M,X)}$ is an H-domain.
Also, since R is not catenary, D is not catenary, by (A.11.1) ⇔
(A.11.19), so there exists P ∈ Spec D such that depth P = 1 and
height P + depth P < altitude D , by (A.9.1) ⇔ (A.9.4). Thus, by
(3.3.5), D is not an H-domain, and this is a contradiction. Therefore R is catenary, so (3.3.5) ⇒ (3.3.6).

Assume (3.3.6) holds, let R be an H-local domain, let R^s be
a special extension of R , and let $L = R^s_N$, where N is a maximal
ideal in R^s . Let p be a height one prime ideal in L , and let
p' be a prime ideal in $L' = (R')_{(R^s-N)}$ such that p'∩L = p , so
height p' = 1 . Now, to prove that L is an H-domain, it may clearly

be assumed that altitude $L > 1$, so depth $p' \geqq 1$, and so depth p'
= altitude $L - 1$, as will now be shown. Namely, by (6.1.1) $\Rightarrow$ (6.1.4)
(and hypothesis), R' satisfies the c.c., so R' is catenary, by
(1.3.5). Also, if M' is a maximal ideal in R' , then height $M' \in$
$\{1$, altitude $R\}$, by (3.2). Thus, since R^s is integral over R ,
$L' = (R')_{(RS-N)}$ is catenary (by (1.2.1)) and, if N' is a maximal
ideal in L' , then height $N' \in \{1$, altitude L = altitude $R\}$. Hence
it follows that height p' + depth p' = altitude L' , so depth p =
depth p' = altitude $L' - 1$ = altitude $L - 1$. Therefore, L is an
H-domain, so (3.3.6) $\Rightarrow$ (3.3.7).

(3.3.7) $\Rightarrow$ (3.3.8), since clearly (3.3.7) $\Rightarrow$ (11.1.8), and (11.1.8)
$\Rightarrow$ (11.1.1) = (3.3.8).

Finally, assume that (3.3.8) holds and let R be a local domain
such that R' satisfies the f.c.c. Then R satisfies the f.c.c., by
(1.2.4), hence, by (3.3.8), R' satisfies the c.c. Therefore, R'
satisfies the c.c. and the f.c.c., hence R' satisfies the s.c.c.,
by (1.3.3), and so R satisfies the s.c.c., by (1.3.4), q.e.d.

As mentioned in the introduction, some comments on where these
conjectures have previously appeared in the literature and some equivalences of each of the named conjectures (and of (3.3.1)) will be given
in Chapters 4 - 13. This is also the case for the other named conjectures that are considered in the remaining theorems in this chapter.

Concerning (3.3.4) $\Rightarrow$ (3.3.6), it was noted in the last paragraph
of Section 4 of [R-12] that (3.4.1) (below) implies the H-Conjecture
holds. Also, two different proofs that the H-Conjecture implies the
Catenary Chain Conjecture were given in [R-5, Remark 3.5(ii)] and [R-6,
(4.5)]. Finally, it was noted without proof in the introduction of
[R-4] that the Catenary Chain Conjecture implies the Normal Chain Conjecture.

The reason for including (3.3.5) is explained following (3.4).

Also, concerning (3.3.7), see the comment preceding (3.10).

(3.4) REMARK. Consider the following statements for a local domain (R,M) :

(3.4.1) If there exists $P \in$ Spec R such that height $P = h$
and depth $P = 1$, then there exists $p \in$ Spec R such that height
$p = 1$ and depth $p = h$.

(3.4.2) (3.4.1) holds for $D = R[X]_{(M,X)}$.

Then (3.3.3) $\Rightarrow$ (3.4.1) $\Rightarrow$ (3.3.4) and (3.4.1) $\Rightarrow$ (3.4.2) $\Rightarrow$ (3.3.5).

Proof. It is clear that (3.4.1) $\Rightarrow$ (3.3.4) and that (3.4.1) $\Rightarrow$
(3.4.2) $\rightarrow$ (3.3.5), so assume that (3.3.3) holds and let (R,M) and P
be as in (3.4.1). If $h \leq 1$, then it is clear that (3.4.1) holds, so
assume that $h > 1$. Let $(0) \subset P_1 \subset \cdots \subset P_h = P \subset M$ be a mcpil h+1
through P , so height $P_i = i$, for $i \leq h$. Then, by (3.3.3) $\Rightarrow$
(5.1.1) $\Rightarrow$ (5.1.3) (with P_{h-2} and P_h the I and P of (5.1.3)),
P_{h-1} can be replaced by $Q_{h-1} \in$ Spec R such that depth $Q_{h-1} = 2$
(and height $Q_{h-1} = h-1$, since $P_{h-2} \subset Q_{h-1} \subset P_h = P$) . Again by
(5.1.1) $\Rightarrow$ (5.1.3) (with P_{h-3} and Q_{h-1} the I and P of (5.1.3)),
P_{h-2} can be replaced by $Q_{h-2} \in$ Spec R such that depth $Q_{h-2} = 3$
(and height $Q_{h-2} = h-2$) . Repetitions of this give a chain of prime
ideals $(0) \subset Q_1 \subset \cdots \subset Q_{h-2} \subset Q_{h-1} \subset P_h = P$ such that depth $Q_{h-i} =$
$i+1$ and height $Q_{h-i} = h-i$. In particular, height $Q_1 = 1$ and
depth $Q_1 = h$, so (3.4.1) holds, q.e.d.

Along with (3.3.3) - (3.3.5) and (3.4), if we consider the statement (*): the Depth Conjecture holds for $D = R[X]_{(M,X)}$; then it is
clear that (3.3.3) $\Rightarrow$ (*), and (*) $\Rightarrow$ (3.4.2) as in the proof that
(3.3.3) $\Rightarrow$ (3.4.1).

The reason for including (3.3.5) in (3.3) is that certain of the
chain conjectures hold for local domains of the form $D = R[X]_{(M,X)}$
(see (14.1)) - and if this can be shown to hold for the Weak Depth

Conjecture (or for (3.4.1) or for the Depth Conjecture), then it follows from (3.3) (or (3.4) or the preceding paragraph) that the H-Conjecture, the Catenary Chain Conjecture, and the Normal Chain Conjecture
also hold.

(3.5) REMARK. (3.5.1) The following statement is stronger than
(3.3.1), and is false: if D is a local domain such that, for all
nonzero P $\in$ Spec D , D/P is catenary, then D is catenary.

(3.5.2) The following statement is weaker than (3.3.9) and is
true: if R is either a Henselian local domain or is of the form
L[X]$_{(M,X)}$ with (L,M) a local domain, and if R' satisfies the
f.c.c., then R satisfies the s.c.c.

Proof. (3.5.1) Let (R,M) be the local domain in [N-6, Example
2, pp. 203-205] in the case m = 0 . Then R is catenary and R'
satisfies the c.c. (since R' is a regular domain with exactly two
maximal ideals - one of height one and the other of height = altitude
R > 1) , hence R satisfies the o.h.c.c. (1.1.7). Therefore, if
there exists a mcpil n in an integral extension domain of R , then
n $\in$ {1 , a = altitude R} , by (A.10.1) $\Rightarrow$ (A.10.11). Hence, if there
exists a mcpil n in D = R[X]$_{(M,X)}$, then n $\in$ {2,a+1} , by (1.5.1) $\Leftrightarrow$
(1.5.4). Therefore it follows that, for each height one prime ideal
P in D , depth P $\in$ {1,a} and D/P is catenary. Hence, for each
nonzero Q $\in$ Spec D , D/Q is catenary (by (1.2.1), since there is a
height one prime ideal P $\subseteq$ Q and then D/Q = (D/P)/(Q/P)) . However,
D is not catenary, since R' has a mcpil 1 , so D has a mcpil 2
< altitude D , by (1.5.1) $\Rightarrow$ (1.5.4).

(3.5.2) If R' satisfies the f.c.c., then R does, by (1.2.4),
so R satisfies the s.c.c., by (A.11.1) $\Leftrightarrow$ (A.11.14) and (A.11.1) $\Leftrightarrow$
(A.11.19), q.e.d.

There is another result that is closely related to (3.3.8) and that is true, namely: if R is a catenary local domain, then R_P satisfies the s.c.c., for all non-maximal $P \in$ Spec R [R-4, Corollary 3.13]. (This is closely related to (3.3.8), because on "inverting" it (see the next paragraph), we get a statement that is equivalent to (3.3.8), namely [R-6, (4.3)]: if R is a catenary local domain, then R/P satisfies the s.c.c., for all nonzero $P \in$ Spec R .

(By "inverting" we mean: interchange "height" and "depth" and interchange "quotient ring" and "factor ring." This sometimes leads to a true result, such as: (1) "a semi-local ring R is analytically unramified if, and only if, R_M is analytically unramified, for all maximal ideals M in R" becomes "a semi-local ring R is analytically unramified if, and only if, R/z is analytically unramified, for all minimal prime ideals z in R" ; (2) "a Noetherian ring A satisfies the f.c.c. if, and only if, for all maximal ideals M in A , A_M satisfies the f.c.c. and height M = altitude A" becomes "a Noetherian ring A satisfies the f.c.c. if, and only if, for each minimal prime ideal z in A , A/z satisfies the f.c.c. and depth z = altitude A" ; and, (3) same as (2) with s.c.c. in place of f.c.c. On the other hand, it sometimes leads to an open problem, such as: (1) the Depth Conjecture (see the introduction to Chapter 5); (2) the H-Conjecture (see the introduction to Chapter 6); and, (3) the equivalence of the Catenary Chain Conjecture mentioned just above. The concept of inverting is a useful one, and I recommend that the reader consider using it after proving any theorem where it is applicable.)

The remaining six theorems in this chapter contain some additional conjectures that are intermediate to the Chain Conjecture and the Normal Chain Conjecture. In each of the theorems, I do not know if any of the reverse implications hold.

The three new statements in (3.6) are concerned with GB-domains, and in (3.6.4) we come to the first non-local domain conjecture. In

(3.6.4) attention could be restricted to the local domain case, by
(7.4.1) ⇒ (7.4.2), but for historical reasons we prefer to state
(3.6.4) for arbitrary Noetherian domains. (Note that (3.6.4) is a
restricted version of W. Krull's question in [Kr, p. 755] - see (2.1).)

(3.6) THEOREM. For the following statements, (3.6.i) ⇒ (3.6.i+1),
for i = 1,2,3,4 :

(3.6.1) The Chain Conjecture (3.3.2) holds.

(3.6.2) If R is a local domain such that R' is level, then
R is a GB-domain (1.1.10).

(3.6.3) The Descended GB-Conjecture holds; that is, if R is
a local domain such that R' is quasi-local, then R is a GB-domain.

(3.6.4) The GB-Conjecture holds; that is, the integral closure
of a Noetherian domain is a GB-domain.

(3.6.5) The Normal Chain Conjecture (3.3.9) holds.

Proof. If (3.6.1) holds and R is a local domain such that R'
is level, then R satisfies the s.c.c., by (4.1.1) ⇔ (4.1.2), so R
is a GB-domain, by (A.11.1) ⇒ (A.11.13), and so (3.6.2) holds.

It is clear that (3.6.2) ⇒ (3.6.3).

Assume (3.6.3) holds and let A be a Noetherian domain. Then A'
is a GB-domain if, and only if, $A'_{M'}$ is, for each maximal ideal M'
in A' , by (A.6.1) ⇔ (A.6.5). Now, if c ∈ M' is such that c is
not in any other maximal ideal in A' that lies over M'∩A , then
$A'_{M'}$ is the integral closure of the local domain $A[c]_{M'∩A[c]}$. Therefore, to prove that (3.6.4) holds, it may be assumed that A is local
and A' is quasi-local, and then A is a GB-domain, by (3.6.3).
Therefore A' is a GB-domeain, by (A.6.1) ⇒ (A.6.2), hence (3.6.4)
holds.

Finally, if (3.6.4) holds and R is a local domain such that R'
satisfies the f.c.c., then R' satisfies the f.c.c. and is a GB-domain,

by (3.6.4), so R' satisfies the s.c.c., by [R-16, (3.10)]. Hence R satisfies the o.c.c., by (1.3.4), and so (3.6.5) holds, q.e.d.

(3.7) PROPOSITION. If (3.6.2) holds, then the Catenary Chain Conjecture (3.3.8) holds.

Proof. Assume (3.6.2) holds and let R be a catenary local domain such that R' is level. Then R is a GB-domain (by (3.6.2)) and satisfies the f.c.c., so R satisfies the s.c.c., by (A.11.1) ⇔ (A.11.13), hence the Catenary Chain Conjecture holds, by (11.1.1) ⇔ (11.1.4), q.e.d.

(3.8) contains three more named conjectures, two of which have previously appeared in the literature under these names, and the third, (3.8.2), is new in these notes.

(3.8) THEOREM. For the following statements, (3.8.i) ⇒ (3.8.i+1), for i = 1,...,5 :

(3.8.1) The Chain Conjecture (3.3.2) holds.

(3.8.2) The Strong Avoidance Conjecture holds; that is, if $P \subset Q \subset N$ is a saturated chain of prime ideals in a semi-local ring R and if $N_1,\ldots,N_h$ are prime ideals in R such that $N \not\subseteq \cup N_i$, then there exists $q \in \operatorname{Spec} R$ such that $P \subset q \subset N$, $q \not\subseteq \cup N_i$, height q = height P + 1 , and depth q = depth N + 1 (so $P \subset q \subset N$ is saturated).

(3.8.3) The Avoidance Conjecture holds; that is, if $P \subset Q \subset N$ is a saturated chain of prime ideals in a Noetherian ring A and if $N_1,\ldots,N_h$ are prime ideals in A such that $N \not\subseteq \cup N_i$, then there exists $q \in \operatorname{Spec} A$ such that $P \subset q \subset N$ is saturated and $q \not\subseteq \cup N_i$.

(3.8.4) If $z \subset p \subset N$ is a maximal chain of prime ideals in a semi-local ring R , then there exists $q \in \operatorname{Spec} R$ such that height q = 1 = depth q .

(3.8.5) The Upper Conjecture holds; that is, if (R,M) is a local

domain <u>and</u> <u>if</u> <u>there</u> <u>exists</u> <u>a</u>   mcpil n + 1 <u>in</u> D = R[X] $_{(M,X)}$, <u>then</u>
<u>either</u> <u>there</u> <u>exists</u> <u>a</u> mcpil n <u>in</u> R <u>or</u> n = 1 .

(3.8.6) <u>The Catenary Chain Conjecture</u> (3.3.8) <u>holds</u>.

<u>Proof</u>. Assume (3.8.1) holds and let R , P ⊂ Q ⊂ N , and N_1,...,
N_h be as in (3.8.2). Then to show the existence of q ∈ Spec R such
that P ⊂ q ⊂ N is saturated, depth q = depth N + 1 , and q ⊄ UN_i ,
it may clearly be assumed that P = (0) . For this case, it will be
shown that there exist infinitely many height one q ∈ Spec R such
that q ⊂ N , depth q = depth N + 1 , and q ⊄ UN_i . Then, since it
is known [Mc-1, Theorem 1] that if P is a prime ideal in a Noetherian
ring A , then at most finitely many q' ∈ Spec A such that P ⊂ q'
and height q'/P = 1 are such that height q' > height P + 1 , it follows that (3.8.2) holds. Therefore we assume to begin with that P =
(0) .

Let S be a finite integral extension domain of R such that
S ⊆ R' and there exists a one-to-one correspondence between the maximal ideals in S and the maximal ideals in R' . Then, for each maximal ideal M in S , S_M satisfies the s.c.c., by (1.3.4), since
$(S_M)' = R'_{(S-M)}$ satisfies the s.c.c. (by hypothesis and (1.3.2)).
Therefore S satisfies the c.c., by (1.3.2), so S is catenary, by
(1.3.5).

Let (0) ⊂ Q' ⊂ N' be a saturated chain of prime ideals in S
that lies over (0) ⊂ Q ⊂ N , so height N' = 2 (since S is catenary). Let I = {N'' ∈ Spec S ; N''∩R ∈ {N_1,...,N_h}} . Assume it is
known that there exist infinitely many height one q' ∈ Spec S such
that q' ⊂ N' and q' ⊄ U = (U{N'' ; N'' ∈ I})∪(U{M ; M is a maximal
ideal in S and N' ⊄ M}) . Then there are infinitely many prime
ideals in {q'∩R ; q' ⊂ N' and q' ⊄ U} , and all but finitely many
of them have height one, by [Mc-1, Theorem 7]. Also, q'∩R ⊂ N and
depth q'∩R = depth q' = depth N' + 1 , since q' is contained in a

maximal ideal M in S only if $N' \subseteq M$ and since S is catenary.
Therefore depth $q' \cap R$ = depth $N' + 1$ = depth $N + 1$. Moreover, $q' \cap R$
$\not\subseteq \cup N_i$, since $q' \not\subseteq \cup \{N'' \ ; \ N'' \in I\}$. Therefore it remains to show the
existence of infinitely many height one $q' \in \operatorname{Spec} S$ such that $q' \subset$
N' and $q' \not\subseteq U$.

For this, since $N' \not\subseteq U$, let $a_1 \in N'$, $\not\in U$. Then there exists
a minimal prime divisor q_1' of $a_1 S$ such that $q_1' \subset N'$. Let a_2
$\in N'$, $\not\in \cup \cup U_1$, where $U_1 = \cup \{q_1' \in \operatorname{Spec} S \ ; \ a_1 \in q_1' \subset N'\}$. Then
there exists a minimal prime divisor q_2' of $a_2 S$ such that $q_2' \subset N'$.
Let $a_3 \in N'$, $\not\in \cup \cup U_1 \cup U_2$, where $U_2 = \cup \{q_2' \in \operatorname{Spec} S \ ; \ a_2 \in q_2' \subset N'\}$.
Repetitions of this show the existence of infinitely many height one
$q' \subset N'$ such that $q' \not\subseteq U$, so (3.8.2) holds.

Assume (3.8.2) holds and let A , $P \subset Q \subset N$, and $N_1, \ldots, N_h$ be
as in (3.8.3). Let $S = A - N \cup N_1 \cup \cdots \cup N_h$, so it suffices to prove that
there exists $q \in \operatorname{Spec} A_S$ such that $P A_S \subset q \subset N A_S$ is saturated and
$q \not\subseteq \cup N_i A_S$, and this follows immediately from (3.8.2).

Assume (3.8.3) holds and let $z \subset p \subset N$ and R be as in (3.8.4).
Then R/z is a semi-local domain and $(0) \subset p/z \subset N/z$ is a mcpil 2
in R/z , so there exists a height one depth one prime ideal in R/z ,
by (3.8.3) = (8.3.1) $\Rightarrow$ (8.3.2). Therefore there exist infinitely many
$q \in \operatorname{Spec} R$ such that $z \subset q$, height $q/z = 1 = $ depth q/z , by (5.2).
Since only finitely many of these q are such that height $q > $ height z
$+ 1 = 1$, by [Mc-1, Theorem 1], (3.8.4) holds.

Assume (3.8.4) holds and let (R, M) and D be as in (3.8.5).
Then, by (3.8.5) = (9.1.1) $\Leftrightarrow$ (9.1.6), it suffices to prove that if
there exists a mcpil $n+1 > 2$ in D , then there exists a mcpil n
in R . For this, let $(0) \subset P_1 \subset \cdots \subset P_{n+1}$ be a mcpil $n+1 > 2$ in
D . Then, by [HMc, Corollary 1.5], there exists a mcpil $n+1$ $(0) \subset$
$Q_1 \subset \cdots \subset Q_{n+1} = P_{n+1}$ in D such that height $Q_i \cap R = i-1$, for
$i = 1, \ldots, n$ (so $Q_i \supset (Q_i \cap R) D$). Let $q = Q_{n-1} \cap R$, so height $q = $

= n-2   and $qD \subset Q_{n-1}$.   Then $(0) \subset Q_{n-1}/qD \subset Q_n/qD \subset Q_{n+1}/qD$ is a

mcpil 3   in $D/qD \cong (R/q)[X]_{(M/q,X)}$, so there exists a finite integral

extension domain S of R/q that has a mcpil 2 , by [HMc, Theorem

1.8].   Therefore, by (3.8.4), there exists a height one depth one prime

ideal in S , so there exists a height one depth one prime ideal in

R/q , by (A.5.5) $\Rightarrow$ (A.5.4), hence there exists a mcpil 2 in R/q .

Therefore, since height q = n-2 , there exists a mcpil n in R , and

so (3.8.5) holds.

Finally, assume (3.8.5) holds and let (R,M) be a catenary local

domain.   Then to prove that (3.8.6) holds it may clearly be assumed

that a = altitude R > 1 . Let P be a height two prime ideal in D =

$R[X]_{(M,X)}$, and let d = depth P , so d > 0 (since a > 1) .   There-

fore there exists a mcpil d+1 in R , by (3.8.5), so d+1 = a , by

hypothesis, and so d = altitude D - 2 . Hence D is an H_2-domain,

so (3.8.6) holds, by (3.8.6) = (11.1.11) $\Rightarrow$ (11.1.1), q.e.d.

The proof of (3.8.1) $\Rightarrow$ (3.8.2) shows that the "there exists" in

(3.8.2) can be replaced by "there are infinitely many." It is shown

in (8.2) that if the Strong Avoidance Conjecture holds, then the "there

exists" in (3.8.2) can be replaced by "there are infinitely many" (with-

out assuming that the Chain Conjecture holds).

The proof of (3.8.4) $\Rightarrow$ (3.8.5) is quite similar to the proof in

[HMc, Proposition 3.8] that the Avoidance Conjecture implies the Upper

Conjecture.

In [HMc, Proposition 3.7], it was shown that the Depth Conjecture

implies the Upper Conjecture (see (3.14.3) below), and in [Mc-1, p. 728]

it was noted without proof that the Avoidance Conjecture implies the

Taut-Level Conjecture (3.9.5) (see (3.14.4) below).

The four new statements in (3.9) are concerned with semi-local

domains.   Only one of these new statements is named, (3.9.5), and it

has previously appeared in the literature under this name.   It should

be noted that (3.9.2) is the semi-local version of the H-Conjecture.
(The semi-local versions of the other previously considered conjectures
for local comains are considered in (3.13), (4.4), (7.3), (9.3), (11.2),
and (12.2).)

(3.9) THEOREM. For the following statements, (3.9.i) ⇒ (3.9.i+1),
for i = 1,...,5 :

(3.9.1) The Strong Avoidance Conjecture (3.8.2) holds.

(3.9.2) If R is an H-semi-local domain, then R satisfies the
f.c.c.

(3.9.3) If R is an H-semi-local domain, then, for each maximal
ideal M in R , R_M is an H-domain.

(3.9.4) If R is a taut semi-local domain, then, for each maximal ideal M in R , R_M is an H-domain.

(3.9.5) The Taut-Level Conjecture holds; that is, if R is a
taut-level semi-local domain, then R satisfies the f.c.c.

(3.9.6) The Catenary Chain Conjecture (3.3.8) holds.

Proof. It is shown in (3.14.1) that (3.8.2) implies the semi-
local Depth Conjecture (3.13.1) holds, and in (3.14.2) it is shown that
the semi-local Depth Conjecture and the Taut-Level Conjecture (3.9.5)
together imply that the semi-local H-Conjecture (3.9.2) holds, so (3.9.1)
⇒ (3.9.2), once we know (3.8.2) ⇒ (3.9.5). For this, let R be a taut-
level semi-local domain and let $(0) \subset Q_1 \subset \cdots \subset Q_n$ be a mcpil n
in R . Then to prove that R satisfies the f.c.c., it suffices to
prove that n = altitude R , and it may clearly be assumed that n > 1 .
Therefore, by (3.8.2) applied to $Q_{n-2} \subset Q_{n-1} \subset Q_n$, it may be assumed
that Q_n is the only maximal ideal in R that contains Q_{n-1} , so
depth $Q_{n-1} = 1$. Also, height $Q_{n-1} = n-1$, by [McR-2, Corollary 8].
Therefore n = height Q_{n-1} + depth Q_{n-1} = altitude R , so (3.9.5) holds,
hence (3.9.1) ⇒ (3.9.2).

It is clear that $(3.9.2) \Rightarrow (3.9.3)$.

Assume $(3.9.3)$ holds and let R be a taut semi-local domain. If either $a = \text{altitude } R = 1$ or there are no height one maximal ideals in R, then R is an H-domain, so each R_M is an H-domain, by $(3.9.3)$. Therefore, assume $a > 1$ and there exists a height one maximal ideal in R. Then, with b and $S = R$ as in (3.1), $B = R[1/b]$ is a taut-level semi-local domain, by $(3.1.1)$ and $(3.1.5)$, so B is an H-domain. Therefore, for each maximal ideal M in R such that height $M > 1$, $R_M = R[1/b]_{MR[1/b]} = B_{MB}$ is an H-domain, by $(3.9.3)$ (and $(3.1.3)$); and it is clear that R_M is an H-domain, if height $M = 1$. Therefore, $(3.9.4)$ holds.

Assume $(3.9.4)$ holds, let R be a taut-level semi-local domain, and let M be a maximal ideal in R. Then, since height $M = \text{alti-}$ tude $R = (\text{say})$ a, to show that R satisfies the f.c.c., it suffices to prove that R_M is catenary, by $(1.2.3)$. For this, it suffices to show that R_M is an H_i-domain, for $i = 1, \ldots, a$, by $(A.9.1) \Leftrightarrow (A.9.5)$. Now R_M is an H-domain, by $(3.9.4)$, so assume R_M is an H_i-domain, for some i $(1 \leq i < a)$. Let $p \in \text{Spec } R$ such that $p \subset M$ and height $p = i$. Then R/p is taut, by [McR-2, Proposition 5], so $R_M/pR_M \cong (R/p)_{M/p}$ is an H-domain, by $(3.9.4)$. Also, altitude $R_M/pR_M = a - i$, since R_M is an H_i-domain, so R_M is an H_{i+1}-domain, by $(A.3.1) \Leftrightarrow (A.3.5)$. Therefore, it follows that R_M is an H_i-domain, for $i = 1, \ldots, a$, hence $(3.9.5)$ holds.

Finally, assume $(3.9.5)$ holds and let A be a taut finite integral extension domain of a local domain R. Then, by hypothesis and $(10.1.1) \Leftrightarrow (10.1.2)$, A is catenary. Therefore $(3.9.6)$ holds, by $(11.1.6) \Rightarrow (11.1.1)$, q.e.d.

In [R-14, (2.14)], it was noted without proof that the Catenary Chain Conjecture holds, if the following condition (formally stronger than $(3.9.4)$) holds: if R is a taut semi-local domain, then, for each maximal ideal M in R , R_M is taut. Also, it was noted without proof in the remark preceding Proposition 9 in [McR 2] that (3.9.5) → (3.9.6).

By (3.9.2) ⇒ (3.9.5), it follows that if R_M is an H-domain whenever R is an H-semi-local domain, then the Taut-Level Conjecture holds. This was a quite unexpected result to me. It certainly seems like the current state of knowledge about semi-local domains should be sufficient to show that either (3.9.2) holds or that it does not hold. However, even when R is an H-semi-local domain that is integral over a local subdomain L , it is not known if each R_M is an H-domain (see (3.10.2)) - and this continues to hold even when R is a special extension of L (by (3.3.7), since (3.3.7) ⇔ (3.10.2)).

The statements in (3.10) are very closely related to those in (3.9).

(3.10) THEOREM. For the following statements, (3.9.2) ⇒ (3.10.1) ⇒ (3.10.2) ⇔ (3.3.7) ⇒ (3.10.3) ⇔ (3.3.8).

(3.10.1) If R is an H-semi-local domain that is integral over a local sub-domain, then R satisfies the f.c.c.

(3.10.2) If R is an H-semi-local domain that is integral over a local sub-domain, then, for each maximal ideal M in R , R_M is an H-domain.

(3.10.3) If R is a taut semi-local domain that is a finite integral extension of a local sub-domain, then R_M is an H-domain.

Proof. It is clear that (3.9.2) ⇒ (3.10.1) ⇒ (3.10.2) ⇒ (3.3.7), so assume that (3.3.7) holds, let R be an H-semi-local domain that is integral over a local subdomain L_0 , and let M be a maximal ideal in R . Let c ∈ M such that 1-c is in all other maximal ideals in R , let $A = L_0[c]$, and let $L = L_0 + J$, where J is the Jacobson radical of A . Then L is a local domain and it is readily seen that A is a special extension of L . Also, L is an H-domain, since R

is integral over L and R is an H-domain. Therefore, by (3.3.7),
$A_{M \cap A}$ is an H-domain. Now R_M is integral over $A_{M \cap A}$, so R is an
H-domain, by (A.3.1) ⇒ (A.3.3), hence (3.10.2) holds.

(3.10.2) ⇒ (3.10.3), as in the proof of (3.9.3) ⇒ (3.9.4) (note
that if R is as in (3.10.3) and L_O is a local domain such that R
is a finite L_O-algebra, then there exists a finite L_O-module L such
that L is local and $L \subseteq R \subseteq L'$, so if b is as in the proof of
(3.9.3) ⇒ (3.9.4), then B = R[1/b] is integrally dependent on L[b,
1/b] and L[b,1/b] is a local domain, by (3.1.2)).

Assume (3.10.3) holds and let A be a taut finite integral extension domain of a local sub-domain. Then, to prove that (3.3.8) holds,
it suffices, by (11.1.6) ⇒ (11.1.1) = (3.3.8), to prove that A is
catenary; that is, by (1.2.2), that A_M satisfies the f.c.c., for all
maximal ideals M in A . Therefore, by (A.9.1) ⇔ (A.9.5), it suffices to prove that each of the rings A_M is an H_i-domain, for i = 1,
..., height M , and the proof of this is similar to the proof of (3.9.4)
⇒ (3.9.5).

Finally, (3.3.8) ⇒ (3.10.3), since (3.3.8) = (11.1.1) ⇒ (11.1.6)
⇒ (3.10.3) (by (1.2.2)), q.e.d.

The next theorem has three new statements, and these are concerned
with Henselizations and with H-local domains.

(3.11) THEOREM. For the following statements, (3.3.2) ⇒ (3.11.1)
⇒ (3.11.2) ⇒ (3.11.3) ⇔ (3.10.2) ⇒ (3.3.8):

(3.11.1) If R is a local domain such that R' is level, then,
for all P ∈ Spec R^H (1.1.14), depth P = depth P∩R .

(3.11.2) If R is an H-local domain, then R^H is an H-ring.

(3.11.3) If R is an H-local domain, then, for all maximal
ideals M' in R' , $R'_{M'}$ is an H-domain.

<u>Proof</u>. Assume (3.3.2) holds and let R be a local domain such that R' is level. Then R satisfies the s.c.c., by (4.1.1) ⇔ (4.1.2), so (3.11.1) holds, by (A.11.1) ⇒ (A.11.15).

Assume (3.11.1) holds and let R be an H-local domain. If a = altitude R = 1 , then it is clear that R^H is an H-ring, so assume a > 1 . If there are no height one maximal ideals in R' , then R' is level, by (3.2), so R^H is an H-ring, by (3.11.1) (since height one prime ideals in R^H lie over height one prime ideals in R) . Therefore, assume there exists a height one maximal ideal in R' and let b , S = R' , and B be as in (3.1), so B = R[b,1/b] is an H-local domain and there are no height one maximal ideals in B' , by (3.1.2), (3.1.4), and (3.1.1). Therefore, by what has already been proved, B^H is an H-ring. Also, by the choice of b and since B' = R'[1/b] , by (3.1.1), $B^H = (R^H/I)[b]$, where I = ∩{z ∈ Spec R^H ; depth z = a} , by [N-6, (43.18)] and its proof together with [N-6, Ex. 2, p. 188]. Therefore, R^H/I is an H-ring, by (A.3.2) ⇒ (A.3.1) (since R^H/I and $(R^H/I)[b]$ have the same total quotient ring). Thus R^H is an H-ring, by the definition of I and since if z is a minimal prime ideal in R^H , then depth z ∈ {height M' ; M' is a maximal ideal in R'} = {1,a} (by [N-6, Ex. 2, p. 188] and (3.2)), hence (3.11.2) holds.

Assume that (3.11.2) holds, let R be an H-local domain, and let M' be a maximal ideal in R' . Then, to show that (3.11.3) holds, it clearly may be assumed that height M' > 1 , so height M' = altitude R, by (3.2). Let c ∈ M' such that c is not in any other maximal ideal in R' , let P = M'∩R[c] , and let C = $R[c]_P$, so C' = $R'_{M'}$. Then there exists a minimal prime ideal z in R^H such that $(R^H/z)[c]$ = C^H (by [N-6, (43.18)] and its proof together with [N-6, Ex. 2, p. 188]). Now, since R^H is an H-ring (by hypothesis), R^H/z is, by (A.3.1) ⇒ (A.3.5), so C^H is, by (A.3.1) ⇒ (A.3.3). Therefore, since C is a dense subspace of C^H , C is an H-domain, by [R-9, (3.11)]. Hence $R'_{M'}$ is an H-domain, by (A.3.1) $\Rightarrow$ A.3.3), and so (3.11.3) holds.

Assume (3.11.3) holds, let R be an H-semi-local domain that is integrally dependent on a local subdomain L_0 , and let M be a maximal ideal in R . Then, to prove that R_M is an H-domain, it may clearly be assumed that R is not local and that a = altitude R > 1 . Let c , A , and L be as in the proof of (3.3.7) $\Rightarrow$ (3.10.2) (in (3.10)), so L is a local domain, R is integral over L and A , and R_M is integral over $L_1 = A_{M \cap A}$. Therefore it suffices to prove that L_1 is an H-domain, by (A.3.1) $\Rightarrow$ (A.3.3). For this, L and A are H-domains (by integral dependence, since R is), so by hypothesis, for each maximal ideal M' in A' = L' , $L'_{M'}$ is an H-domain. Also, height M' $\in$ {1,a} , by (3.2). Therefore, let p be a height one prime ideal in A such that $p \subset N = M \cap A$ (height N > 1 , since A is an H-domain and a > 1) . Then there exist prime ideals $p' \subset N'$ in A' = L' such that $p' \cap A = p$ and $N' \cap A = N$, so height N' = a . Hence height N/p $\cong$ height N'/p' = height N' - 1 (since $L'_{N'}$ is an H-domain) = altitude L' - 1 = altitude L - 1 , so it follows that $L_1 = A_N$ is an H-domain, and so (3.10.2) holds.

Finally, since it has already been shown that (3.10.2) $\Rightarrow$ (3.3.8), it remains to show that (3.10.2) $\Rightarrow$ (3.11.3). For this, let R be an H-local domain and let M' be a maximal ideal in R' , so it must be shown that $R'_{M'}$ is an H-domain. For this, it may clearly be assumed that height M' > 1 . Let c $\in$ M' such that c is not in any other maximal ideal in R' , let A = R[c] , and let N = M' $\cap$ A , so $(A_N)' = R'_{M'}$. Now, if there are no height one maximal ideals in R' , then R' is an H-domain, by (3.2), so A is, by integral dependence, hence A_N is, by (3.10.2), and so $R'_{M'} = (A_N)'$ is an H-domain, by (A.3.1) $\Rightarrow$ (A.3.3). Therefore assume that R' has a height one maximal ideal and let b , S = R' , and B be as in (3.1). Then B = R[b,1/b] is an H-local domain and there are no height one maximal ideals in $B' = R'[1/b]$, by (3.1.2), (3.1.4), and (3.1.1). Therefore, since $B \subseteq A[b,1/b] \subseteq B' = R'[1/b]$ and $R'_{M'} = R'[1/b]_{M'R'[1/b]}$ is integral over $A[b,1/b]_{N'}$, where $N' = M'R'_M \cap A[b,1/b]$, it follows from what has already been proved that $R'_{M'}$ is an H-domain, and so (3.11.3) holds, q.e.d.

(3.12) REMARK. By (6.1.1) $\Rightarrow$ (6.1.9), the H-Conjecture implies (3.11.2).

In (3.9.2) the semi-local version of the H-Conjecture was given. In (3.13) we make some comments concerning the semi-local version of this conjecture, of the Depth Conjecture, and of the Weak Depth Conjecture. (The semi-local versions of the other previously considered conjectures for a local domain are equivalent to the local versions, as will be shown in (4.4), (7.3), (9.3), (11.2), and (12.2).) These three conjectures lie between the Strong Avoidance Conjecture and the H-Conjecture, as follows from (3.9), (3.13), (3.14.1), and (3.3).

(3.13) REMARK. Together with (3.9.2), consider the following statements:

(3.13.1) If R is a semi-local domain and $P \in$ Spec R is such that height $P > 1$, then there exists $p \in$ Spec R such that $p \subset P$ and depth $p =$ depth $P + 1$.

(3.13.2) If R is a semi-local domain and $P \in$ Spec R is such that height $P = h$ and depth $P = 1$, then there exists $p \in$ Spec R such that height $p = 1$ and depth $p \leq h$.

Then it is clear that (3.9.2) $\Rightarrow$ (3.3.6), (3.9.2) $\Rightarrow$ (3.9.3), (3.13.1) $\Rightarrow$ (3.3.3), and (3.13.2) $\Rightarrow$ (3.3.4); and (3.13.1) $\Rightarrow$ (3.13.2), as in the proof that (3.3.3) $\Rightarrow$ (3.3.4). Also, (3.9.2) is equivalent to "(3.3.6) and (3.9.3)." (For, if R is an H-semi-local domain, then, for each maximal ideal M in R , R_M is an H-domain, by (3.9.3), so R_M satisfies the f.c.c., by (3.3.6); and it is readily seen that R is level (since R is semi-local and is an H-domain), so R satisfies the f.c.c., by (1.2.3).) Moreover, it was shown in (3.3) that (3.3.3) $\Rightarrow$ (3.3.4) $\Rightarrow$ (3.3.6), but for the semi-local case I have not even been able to prove that (3.13.1) $\Rightarrow$ (3.9.2).

The final theorem in this chapter gives four more implications between the named conjectures that have already been considered.

(3.14) THEOREM. The following statements hold:

(3.14.1) If the Strong Avoidance Conjecture (3.8.2) holds, then the semi-local Depth Conjecture (3.13.1) holds.

(3.14.2) If the Taut-Level Conjecture (3.9.5) and the semi-local Depth Conjecture (3.13.1) hold, then the semi-local H-Conjecture (3.9.2) holds,

(3.14.3) [HMc, Proposition 3.7]. If the Depth Conjecture (3.3.3) holds, then the Upper Conjecture (3.8.5) holds.

(3.14.4) [Mc-1, p. 728]. If the Avoidance Conjecture (3.8.3) holds, then the Taut-Level Conjecture (3.9.5) holds.

Proof. (3.14.1) Assume (3.8.2) holds and let P be a prime ideal in a semi-local domain R such that height P > 1 . Let q , q' $\in$ Spec R such that q $\subset$ q' $\subset$ P and height P/q = 2 . Then it readily follows from (3.8.2) that there exists p $\in$ Spec R such that p $\subset$ P and depth p = depth P + 1 , so (3.8.2) $\Rightarrow$ (3.13.1).

(3.14.2) Assume (3.9.5) and (3.13.1) hold and let R be an H-semi-local domain. Suppose there exists P $\in$ Spec R such that height P = h , depth P = d , and h+d < a . Then h > 1 , since R is an H-domain, so by (3.13.1), there exists P_1 $\in$ Spec R such that $P_1 \subset P$ (so height $P_1 \leqq$ h-1) and depth P_1 = d+1 . Repetitions of this show that there exists a chain of prime ideals $P_{k-1} \subset \cdots \subset P_1 \subset P$ (k $\leqq$ h) in R such that height $P_{i+1} \leqq$ height P_i - 1 , depth P_i+1 = depth P_i

+ 1 = d+i+1 , and height P_{k-1} = 1 . In particular, depth P_{k-1} = d+k-1 . Therefore a-1 = depth P_{k-1} = d+k-1 ≦ d+h-1 < a-1 , and this is a contradiction. Thus R is taut-level, so R satisfies the f.c.c., by (3.9.5), and so (3.9.2) holds.

(3.14.3) Assume (3.3.3) holds, let (R,M) be a local domain, and assume there exists a mcpil n+1 > 2 in D = $R[X]_{(M,X)}$, say (0) ⊂ P_1 ⊂ ⋯ ⊂ P_{n+1} = (M,X)D . Then, by [HMc, Proposition 3.6] (see (5.4)), it may be assumed that height P_1 = 1 and depth P_1 = n . Let p = $P_1 \cap R$, so height p ∈ {0,1} . If height p = 1 , then P_1 = pD , so depth p = depth pD - 1 = depth P_1 - 1 = n-1 , and so there exists a mcpil n in R . On the other hand, if height p = 0 , then [Mc-1, Proposition 1 (iii) ⇒ (i)] says that there exists a height one prime ideal q in R such that depth q = n-1 , so again there exists a mcpil in R , and so (3.8.5) holds, by (9.1.1) ⇔ (9.1.6).

(3.14.4) was essentially proved in the last half of the proof of (3.9.1) ⇒ (3.9.2), since (3.8.3) applied to Q_{n-2} ⊂ Q_{n-1} ⊂ Q_n also shows that it may be assumed that Q_n is the only maximal ideal in R that contains Q_{n-1} , q.e.d.

The proof of (3.14.3) is the same as that in [HMc, Proposition 3.7].

A diagram of the implications between the named conjectures that have been proved in this chapter is given on the next page. In the diagram, the numbers under a conjecture indicate where it is stated, and the numbers on the lines between the conjectures indicate where the implication is proved.

44

DIAGRAM OF MAJOR IMPLICATIONS

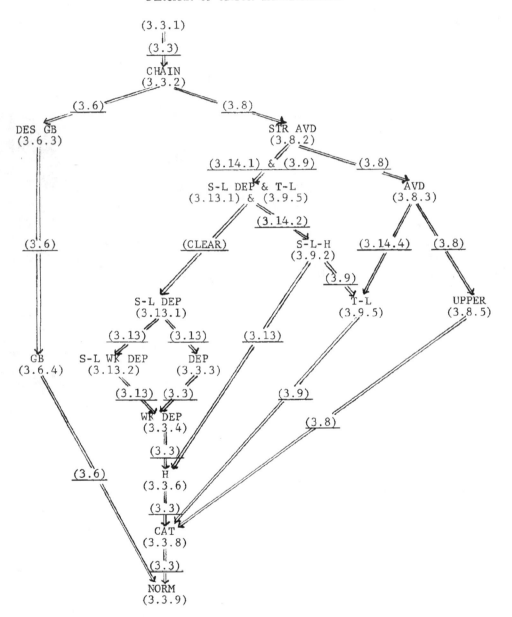

CHAPTER 4

THE CHAIN CONJECTURE

The main result in this chapter, (4.1), gives nine statements that
are equivalent to the Chain Conjecture. But, before giving these
equivalences, a brief history of this conjecture will be given.

Since M. Nagata's example showing that the answer to the chain
problem of prime ideals is no was not integrally closed (see the introduction to Chapter 2 and (B.1)), a natural follow-up question is if
every integrally closed local domain is catenary, and the Chain Conjecture is a generalization of this question (since, by (4.5), the Chain
Conjecture is equivalent to: the integral closure of a local domain
is catenary).

This conjecture was (essentially) given by M. Nagata in 1956, in
[N-3, Problem 3", p. 62]. In [N-3, Problems 3 and 3', p. 62], two
equivalences of the conjecture were given (see (4.1.1) ⟺ (4.1.2) ⟺
(4.1.3)), and, as mentioned in (2.3), it was indicated in [N-5] that
this conjecture holds, if the answer to (2.3)(*) is no. (But the answer
is yes, as was noted in (2.3).) Also, in [N-3, Problem 4, p. 62], the
following result was mentioned without proof: if (H,M) is a Henselian
local domain and u is an element in the quotient field of H such
that u and 1/u are not in H' , then altitude $H[u]_{MH[u]}$ = altitude
H - 1 . This is equivalent to saying that every Henselian local domain
is an H-domain, by (A.3.1) ⟺ (A.3.7) (and (A.2)), and this, in turn,
is equivalent to saying that the Chain Conjecture holds, by (4.1.1) ⟺
(4.1.4).

Next, A. Grothendieck asked the following question in 1965, in
[G-2, p. 103]: if in every finite integral extension domain of a local
domain R , all maximal ideals have the same height, then does R satisfy the s.c.c.? This is equivalent to the Chain Conjecture, as follows
from (4.1.1) ⟺ (4.1.2). Also, in 1967, in [G-3, (18.9.6)(ii)], he asked

if every Henselian local domain is quasi-unmixed; and this is equivalent to the Chain Conjecture, by (4.1.1) ⟺ (4.1.2) and (A.11.1) ⟺
(A.11.4).

Then, at the end of his 1972 paper [K], I. Kaplansky asked the
weaker question mentioned above: is an integrally closed Noetherian
domain necessarily catenary?

In 1973, in [R-11, (2.20)], I showed that the Chain Conjecture
holds for level Noetherian Hilbert domains, if the following condition
is satisfied: if D is a level Noetherian Hilbert domain, then D[b]
is level, for all b in the quotient field of D . But this condition
does not hold, as follows from K. Fujita's result mentioned in (2.8).

Finally, the Chain Conjecture does hold for a large class of local
domains (the local domains that satisfy the s.c.c., (for example, complete local domains, regular local rings, and homomorphic images of
Macaulay rings [N-6, (34.8)]), by (A.11.1) ⟹ (A.11.7)), but it is known
that it does not hold for at least some quasi-local domains. Namely,
J. Sally showed in 1970, in [S], that there exists an integrally closed
quasi-local domain that is not catenary, and I. Kaplansky's example mentioned in (2.1) also showed this.

In (4.1), a number of equivalences of the Chain Conjecture are
given, and among these are that certain local domains are either H-
domains or have a mcpil n or satisfy the f.c.c. or the s.c.c. Now
there are a great many ways of saying that a local domain satisfies
one of these conditions (see (A.3), (A.5), (A.9), and (A.11)), and
each of these ways gives rise to a formally different equivalence of
the Chain Conjecture. Since some of the characterizations of a number
of the other conjectures considered in Chapter 3 also involve these
three conditions, as well as other conditions (such as C_i , GB ,
o.h.c.c., etc.), it was decided to list in Appendix A some of the equivalences of these four conditions (and also of: an H_i-local ring (A.3);

a C_i-local ring (A.4); a GB-ring (A.6); a taut local ring (A.7); a taut semi-local domain (A.8); and, the o.h.c.c. (A.10)), and then make a corresponding reduction in the number of equivalences that are given in Chapters 4 - 13 of the Chain Conjecture and the other conjectures. Thus, with (A.3), (A.5), (A.9), and (A.11) in mind, we now given nine equivalences of the Chain Conjecture. (We restate the Chain Conjecture in (4.1.1) for the reader's convenience. This will also be done for the other named conjectures in Chapters 5 - 13.)

(4.1) THEOREM. The following statements are equivalent:

(4.1.1) The Chain Conjecture (3.3.2) holds: the integral closure of a local domain satisfies the c.c.

(4.1.2) If R is either a Henselian local domain or a local domain such that R' is level, then R satisfies the s.c.c.

(4.1.3) If R is as in (4.1.2), then R satisfies the f.c.c.

(4.1.4) If R is as in (4.1.2), then R is an H-domain.

(4.1.5) If R is as in (4.1.2), then R^* (1.1.14) is an H-ring.

(4.1.6) If R is a local domain and there exists a depth n minimal prime ideal in R^* , then there exists a height n maximal ideal in R' .

(4.1.7) If R is a local domain and there exists a mcpil n in R , then there exists a height n maximal ideal in R' .

(4.1.8) If R is a local domain with quotient field F and $u \in F$ is such that altitude R[u] < altitude R , then $1/u \in M'R'_{M'}$, for all maximal ideals M' in R' such that height M' = altitude R .

(4.1.9) If R is as in (4.1.2), then $D = R[X]_{(M,X)}$ is either an H-domain or an H_2-domain.

(4.1.10) If (R,M) is a local domain and there exists a mcpil n+1 in $D = R[X]_{(M,X)}$, then there exists a height one depth n prime ideal in D that contains a monic polynomial.

Proof.   Assume (4.1.1) holds and let   R   be a local domain such that   R'   is level.   (If   R   is Henselian, then   R'   is quasi-local, hence   R'   is level.)   Then   R'   is level and satisfies the c.c., so   R'   satisfies the s.c.c., by (1.3.3), hence   R   satisfies the s.c.c., by (1.3.4), and so (4.1.1) $\Rightarrow$ (4.1.2).

(4.1.2) $\Rightarrow$ (4.1.3), by (1.3.5), and it is clear that $(4.1.3) \Rightarrow$ (4.1.4).

Assume that (4.1.4) holds.   Then   R   is an   H-domain whenever   R   is a Henselian local domain, and it is known [R-6, (2.4)] that this implies that (4.1.1) holds.

(4.1.2) $\Rightarrow$ (4.1.5), by (A.11.1) $\Leftrightarrow$ (A.11.16), and (4.1.5) $\Rightarrow$ (4.1.2), as will now be shown.   Namely, let   R   be as in (4.1.2).   Then it may clearly be assumed that a = altitude $R > 1$.   If z is a minimal prime ideal in R^* , then   depth $z > 1$ (by [R-2, Proposition 3.5], since there are no height one maximal ideals in   R' , by hypothesis).   Thus, since only finitely many $p \in$ Spec R^* such that $z \subset p$ and   height $p/z = 1$ have   height $>$ height $z + 1 = 1$, by [Mc-1, Theorem 1], there exists a height one prime ideal p in R^* such that $z \subset p$, hence depth z = depth $p + 1 = a$, by (4.1.5).   Therefore   R   is quasi-unmixed, so (4.1.2) holds, by (A.11.4) $\Rightarrow$ (A.11.1).

Assume (4.1.1) holds, let   R   be a local domain, and assume there exists a   depth n minimal prime ideal in R^* .   Then there exists an integral extension domain   A   of   R   that has a   mcpil n , by (1.5.2) $\Rightarrow$ (1.5.1), so   A[R']   is an integral extension domain of   R'   that has a   mcpil n , by the Going Up Theorem.   Therefore, since   R'   satisfies the c.c., there exists a   height n maximal ideal in   R' , so (4.1.6) holds.

(4.1.6) implies that a Henselian local domain is quasi-unmixed, so (4.1.6) $\Rightarrow$ (4.1.2), by (11.1.4) $\Rightarrow$ (11.1.1).

Assume (4.1.6) holds, let   R   be a local domain, and assume there exists a mcpil n in R . Then there exists a depth n minimal prime ideal in R^* , by (1.5.1) → (1.5.2). Therefore there exists a height n maximal ideal in R' , by (4.1.6), and so (4.1.7) holds.

(4.1.7) implies all Henselian local domains satisfy the f.c.c., so (4.1.7) ⇒ (4.1.3).

Assume (4.1.1) holds, let R , F , and u be as in (4.1.8), and suppose that $1/u \not\in M'R'_{M'}$, for some maximal ideal M' in R' such that height M' = altitude R . Then $1 \not\in M'R'[u]$, by [ZS-2, p. 325], so there exists a maximal ideal N in R'[u] such that N∩R' = M' . Now, there is a finite integral extension domain A of R such that $A \subseteq R'$ and such that there is a one-to-one correspondence between the maximal ideals in A and in R' . Therefore, since R' satisfies the c.c. (by hypothesis), [R-2, Theorem 3.10] says that R' satisfies the altitude formula. Thus height N + trd (R'[u]/N)/(R'/M') = height M' + trd R'[u]/R' . That is, since N is maximal, height N = height M' (= altitude R) . Therefore altitude R[u] = altitude R'[u] ≥ altitude R , and this is a contradiction, so (4.1.8) holds.

Assume (4.1.8) holds and suppose that the Henselian part of (4.1.4) does not hold. Then there exists a Henselian local domain (R,M) and u ∈ F such that u and 1/u are not in R' and altitude R[u] < altitude R , by (A.3.1) ↔ (A.3.9) (for i = 1 and see (A.2)), and this contradicts (4.1.8). Therefore (4.1.8) ⇒ (4.1.4).

It is known (A.4.1) ↔ (A.4.7) that D in (4.1.9) is an H_i-domain if, and only if, R is a C_{i-1}-domain. Therefore, (4.1.9) ⇒ (4.1.4) and (4.1.2) ⇒ (4.1.9).

Assume (4.1.6) holds and let (R,M) be a local domain such that there exists a mcpil n+1 in $D = R[X]_{(M,X)}$. Then there exists a depth n minimal prime ideal in R^*, by (1.5.4) ⇒ (1.5.2), so, by (4.1.6), there exists a height n maximal ideal M' in R' . Therefore, with c ∈ M' such that c is not in any other maximal ideal in R' , B =

$R[c]_{M'\cap R[c]}$ has altitude = n . Hence, since $M'\cap R[c] = (M,c)R[c]$, there exists a height one depth n prime ideal K in D such that K contains a monic polynomial and $B = D/K$. Thus (4.1.6) ⇒ (4.1.10).

Finally, assume (4.1.10) holds, let R be a local domain, and assume there exists a depth n minimal prime ideal in R^* . Then there exists a mcpil $n + 1$ in D , by (1.5.2) ⇒ (1.5.4), so it follows from (4.1.10) that there exists a principal integral extension domain $A = R[c]$ of R that has a maximal ideal of height = n . Therefore there exists a height n maximal ideal in R' , by [R-4, Lemma 2.9], hence (4.1.6) holds, q.e.d.

Another proof that (4.1.1) ⟺ (4.1.4) was given in [R-21, (13.6) and (13.7)]. Also, the proof of (4.1.1) ⇒ (4.1.6) ⇒ (4.1.2) is essentially the same as that given in [RMc, (2.4)] to prove that (4.1.6) is equivalent to the Chain Conjecture.

(4.2) REMARK. Two additional characterizations of the Chain Conjecture are known, namely:

(4.2.1) For each maximal ideal M' in the integral closure R' of a local domain R , every DVR (discrete valuation ring) (V,N) in the quotient field of R such that: $R' \subseteq V$; $N\cap R' = M'$; and, V is integral over a locality over R ; is of the first kind (that is, trd $(V/N)/(R'/M')$ = height M' - 1) , [R-13, (2.19.1)]; and,

(4.2.2) If R is a local domain such that R' is level, then $R' = \cap\{V ; (V,N) \in \mathcal{v}\}$, where $\mathcal{v} = \{(V,N) ; V$ is a valuation ring in the quotient field of R , $R \subseteq V$, $N\cap R = M$, and altitude V = altitude R} [R-21, (13.6)].

(It is shown in [R-21] that $\cap\{V ; (V,N) \in \mathcal{v}\}$ (with $\mathcal{v}$ as in (4.2.2)) has many properties that are analogous to the known properties of R' .)

(4.3) contains some comments on the statements in (4.1).

(4.3)   REMARK.   (4.3.1)   (4.1.6) is equivalent to: if there exists a   depth n   minimal prime ideal in R^* , then there exists a depth n   minimal prime ideal in R^H .

(4.3.2)   (4.1.6) holds for   n = 1 .

(4.3.3)   (3.4.2) together with the condition that some such height one prime ideal contains a monic polynomial is equivalent to (3.3.2).

(4.3.4)   The condition in (4.1.10) holds for complete local rings.

Proof.   (4.3.1) follows from [N-6, Ex. 2, p. 188] and the fact that R^H is a dense subspace of R^* .

(4.3.2) is given in [R-2, Proposition 3.5].

(4.3.3) It is known (A.5.1) ⇔ (A.5.4) that if   (R,M)   is a local domain, then there exists a   mcpil n+1   in $D = R[X]_{(M,X)}$ if, and only if, there exists a   height n   depth one prime ideal in   D .   Therefore (4.3.3) follows from (4.1.10) ⇔ (4.1.1).

(4.3.4)   Let   (R,M)   be a complete local ring and assume there exists a   mcpil n+1   in $D = R[X]_{(M,X)}$.   Then there exists   P ∈ Spec D   such that   height P = n   and   depth P = 1 , by (A.5.1) ⇒ (A.5.4).   Let z   be a minimal prime ideal in   D   such that   z ⊆ P   and   height P/z = n .   Then, since   R   is complete,   depth z = n+1 .   Therefore   p = (z,X)D   is a height one prime ideal (since   p/XD = z∩R   is a minimal prime ideal), and   depth p = n   (since   D/p = R/(z∩R)   and   depth z = depth z∩R + 1) , q.e.d.

The final theorem in this chapter shows that this conjecture could just as well have been stated for an arbitrary Noetherian domain.

(4.4)   THEOREM.   The following statements are equivalent:

(4.4.1)   The Chain Conjecture (4.1.1) holds.

(4.4.2)   The integral closure of a Noetherian domain is catenary.

(4.4.3)   The integral closure of a Noetherian domain satisfies the c.c.

Proof. (4.4.3) ⇒ (4.4.2), by (1.3.5).

Assume (4.4.2) holds and let R be a local domain such that R' is level.   Then R' satisfies the f.c.c., by hypothesis and (1.2.3), so R satisfies the f.c.c., by (1.2.4), hence (4.4.1) holds, by (4.1.3) ⇒ (4.1.1).

Finally, assume (4.4.1) holds and let A be a Noetherian domain. Then to prove that A' satisfies the c.c., it suffices, by (1.3.2), to prove that if M' is a maximal ideal in A' , then $A'_{M'}$ satisfies the c.c.   For this, $A'_{M'}$ is the integral closure of a local domain (as in the proof of (3.6.3) ⇒ (3.6.4)), so $A'_{M'}$ satisfies the c.c., by hypothesis, and so (4.4.3) holds, q.e.d.

(4.5)   REMARK.   The Chain Conjecture (4.1.1) is also equivalent to (*):   the integral closure of a local domain is catenary.

Proof. (4.1.1) ⇒ (*), by (1.3.5), and (*) ⇒ (4.1.1), by the proof of (4.4.2) ⇒ (4.4.1), q.e.d.

CHAPTER 5

THE DEPTH CONJECTURE AND THE WEAK
DEPTH CONJECTURE

The Depth Conjecture arose from trying to "invert" (see the discussion between (3.5) and (3.6)) the following result [Mc-1, Theorem
1]: if R is a local domain and P ∈ Spec R is such that depth
P > 1 , then all but finitely many Q ∈ Spec R such that P ⊂ Q and
and height Q/P = 1 are such that height Q = height P + 1 . In
(B.5.8) we give an example to show that the result does not invert, so
the Depth Conjecture arose from a natural follow-up question.

I stated the Depth Conjecture in 1972, in a preliminary version
of these notes, and therein it was shown that the Chain Conjecture ⇒
the Depth Conjecture ⇒ the H-Conjecture. (This fact was briefly commented on in [P, p. 72].) S. McAdam mentioned the Depth Conjecture in
1974, in [Mc-1, Remark, p. 720], and he then gave an example similar
to (but somewhat more complicated than) the one in (B.5.7). Then, in
1975, in [HMc, Proposition 3.7], he and E. Houston showed that this
conjecture implies the Upper Conjecture.

The Weak Depth Conjecture is new in these notes, but as mentioned
after the proof of (3.3), it was noted in [R-12] that the formally
stronger (3.4.1) implies the H-Conjecture.

Related to these two depth conjectures, in 1971, in [R-4, Corollary
2.4(2)], I showed that if a Noetherian domain A is such that altitude A = a < ∞ and A is a D_i-<u>domain</u> (that is, for all P ∈ Spec A
such that depth P = i , height P = a - i) , then A is a D_j-domain,
for j = i , i + 1,...,a . Also, in 1976, in [Hou-1, Theorem 12], E.
Houston showed that a local domain (R,M) is a D_i-domain if, and only
if, $R[X]_{(M,X)}$ is a D_{i+1}-domain.

Finally, the Depth Conjecture and the Weak Depth Conjecture hold
for all catenary local domains, but they do not hold for all Noetherian

domains or quasi-local domains. For example, in [Fu-1, Proposition, p. 482], K. Fujita gave an example of a Noetherian domain with infinitely many maximal ideals that does not satisfy the Depth Conjecture. And, in (14.7), it is shown that both the Depth Conjecture and the Weak Depth Conjecture fail to hold for some non-Noetherian quasi-local domains.

(5.1) gives two equivalences of the Depth Conjecture.

(5.1) THEOREM. The following statements are equivalent:

(5.1.1) The Depth Conjecture (3.3.3) holds: if R is a local domain and $P \in$ Spec R is such that height $P > 1$, then there exists $p \in$ Spec R such that $p \subset P$ and depth p = depth $P + 1$.

(5.1.2) If P is a height two prime ideal in a local domain R , then there exists $p \in$ Spec R such that $p \subset P$ and depth p = depth $P + 1$.

(5.1.3) If R is a local ring, $P \in$ Spec R , and I is an ideal in R such that $I \subset P$ and height $P/I > 1$, then there exist infinitely many $p \in$ Spec R such that $I \subset p \subset P$ and depth p = depth P $+ 1$.

Proof. $(5.1.3) \Rightarrow (5.1.1)$ (with I = (0)) , and it is clear that $(5.1.1) \Rightarrow (5.1.2)$, so assume that (5.1.2) holds and let R , P , and I be as in (5.1.3). Let $q \in$ Spec R such that $I \subseteq q \subset P$ and height $P/q = 2$. Then, by (5.1.2), there exists $p \in$ Spec R such that $q \subset p \subset P$ and depth p = depth $P + 1$. Therefore, by (5.2), there exist infinitely many such p , so (5.1.3) holds, q.e.d.

(5.2) has already been used in the proof of $(3.8.3) \Rightarrow (3.8.4)$ and of (5.1), and it will be used in the proof of (5.3) and in Chapter 8. It follows quite readily from two results in [Mc-1], but is important enough to be explicitly stated.

(5.2) PROPOSITION. Let $P \subset Q \subset N$ be a saturated chain of prime ideals in a semi-local ring S such that depth Q - depth N | 1 . Then there exist infinitely many $q \in$ Spec S such that $P \subset q \subset N$, height q = height P + 1 , and depth q = depth N + 1 (so $P \subset q \subset N$ is saturated).

Proof. $P/P \subset Q/P \subset N/P$ is a saturated chain of prime ideals in S/P , so there exist infinitely many height one prime ideals $q/P \subset$ N/P in S/P such that depth q/P = depth N/P + 1 , by [Mc-1, Proposition 2]. Therefore there exist infinitely many $q \in$ Spec S such that $P \subset q \subset N$ is saturated and depth q = depth N + 1 , so the conclusion follows since only finitely many of these q are such that height q > height P + 1 , by [Mc-1, Theorem 1], q.e.d.

The final theorem in this chapter sharpens [HMc, Proposition 3.6] and shows that the Weak Depth Conjecture is equivalent to a considerably stronger looking result.

(5.3) THEOREM. The following statements are equivalent:

(5.3.1) The Weak Depth Conjecture (3.3.4) holds: if R is a local domain and $P \in$ Spec R is such that height P = h and depth P = 1 , then there exists $p \in$ Spec R such that height p = 1 and depth p $\leq$ h .

(5.3.2) If (R,M) is a local domain and $P \in$ Spec R is such that height P = h and depth P = d , then there exist $n \leq h + d$ and a mcpil n , say $(0) \subset P_1 \subset \cdots \subset P_n = M$, in R such that height P_i = i and depth P_i = n - i , for i = 1,...,n - 1 . In fact, for each i = 1, ...,n - 1 , there are infinitely many choices for P_i with the remainder of the chain unchanged.

Proof. It is clear that (5.3.2) $\Rightarrow$ (5.3.1), so assume (5.3.1) holds and let (R,M) and P be as in (5.3.2). Then there exists $Q \in$ Spec R such that height $Q = h + d - 1$ and depth $Q = 1$ (this is clear if $d = 0$ or $d = 1$, and it follows from (A.5.1) $\Rightarrow$ (A.5.4), if $d > 1$), so there exist $n \leq h + d$ and $P_1 \in \text{Spec } R$ such that height $P_1 = 1$ and depth $P_1 = n - 1$, by (5.3.1). Let $(0) = P_0 \subset P_1 \subset P_2 \subset \cdots \subset P_{n-1} \subset P_n = M$ be a mcpil n through P_1, so depth $P_i = n - i$, for $i = 1, \ldots, n$. Assume that for some i ($1 \leq i \leq n - 1$) height $P_{i-1} = i - 1$. (This holds for $i = 1$.) Then, by (5.2) applied to $P_{i-1} \subset P_i \subset P_{i+1}$, there exist infinitely many $p_i \in \text{Spec } R$ such that $P_{i-1} \subset p_i \subset P_{i+1}$ is saturated, height $p_i = i$, and depth $p_i = \text{depth } P_{i+1} + 1 = n - i$. The conclusion readily follows from this, q.e.d.

(5.4) REMARK. The analogous result to (5.3) for the Depth Conjecture was given in [HMc, Proposition 3.6]: given a mcpil n up to a prime ideal P in a Noetherian domain R, there exists a maximal chain of prime ideals $(0) \subset P_1 \subset \cdots \subset P_n = P$ in R such that height $P_i = i$ and height $P/P_i = n - i$, for $i = 1, \ldots, n - 1$, if the Depth Conjecture holds.

CHAPTER 6

THE H-CONJECTURE

This conjecture, like the Depth Conjecture, arose from trying to
"invert" a known result, namely [R-5, Remark 2.6(i)]: if R is a local domain and all depth one P $\in$ Spec R are such that height P =
altitude R - 1 , then R is catenary.

The conjecture was first stated in 1971, in [R-4, p. 1096] (and
also in [R-4, Remark 2.5(b)] in a different form), and it was pointed
out in 1972, in [R-6, p. 225], that the reason "H" was used is that
it is hoped that all Henselian local domains are H-domains (that is,
by (4.1.1) $\Leftrightarrow$ (4.1.4), that the Chain Conjecture holds).

As mentioned after the proof of (3.3), two different proofs that
the H-Conjecture implies the Catenary Chain Conjecture were given in
[R-5, Remark 3.5(ii)] and [R-6, (4.5)], and in [R-12, p. 130] it was
shown that (3.4.1) implies the H-Conjecture. Also, in [R-21, (13.8)
and (13.10)], it was shown that the H-Conjecture holds, if, for all
local domains (R,M) such that R' is level, $\cap\{V ; (V,N) \in \upsilon\}$ (where
υ is as in (4.2.2)) either satisfies the f.c.c. or is taut. Moreover,
it was noted without proof in the comment following (5.7) in [R-21]
that (6.1.5) is equivalent to this conjecture.

Finally, it is known that this conjecture does not hold for some
nonlocal domains. For example, W. Heinzer's example, (14.6), shows that
there exist H-quasi-local domains that are not carenary. Also, K.
Fujita in [Fu-1, Proposition, p. 484] gave a different example of an
H-quasi-local domain that is not catenary, and in [Fu-1, Proposition,
p. 482] he gave an example of an H-Noetherian domain that is not taut.
And, another example of a noncatenary H-quasi-local domain is given in
(14.7).

The main result in this chapter, (6.1), gives eleven equivalences

of the H-Conjecture. Among these are that certain local rings are
H_i-local rings, or are taut, or satisfy the f.c.c., the o.h.c.c., or
the s.c.c., so we remind the reader that a number of equivalences of
these conditions are listed in (A.3) and (A.7) - (A.11).

The list of equivalences of the H-Conjecture given in (6.1) is
very similar to the list of equivalences of the Catenary Chain Conjecture given in (11.1). Since the lists are given in approximately the
same order, the reader should have no difficulty in matching the corresponding characterizations. However, two comments should be made.
First, it follows immediately from (A.10.1) ⇔ (A.10.12) that (6.1.4) ⇔
(6.1.9). The reason for including this equivalence is that in comparing the conclusions of the corresponding statements in (6.1) and (11.1),
it seems like it should be possible to replace the conclusion of (6.1.9)
with: R^H is an H-ring; but I have been unable to prove this. (Concerning this, see (3.11) and (3.12).) The second comment is that in one
case, the corresponding statements have quite different numbers. Namely, (6.1.7) corresponds to (11.1.12), since, by (A.4.1) ⇔ (A.4.7), the
conclusion of (11.1.12) is equivalent to: R/p is a C_o-domain, for
all height one $p \in$ Spec R ; and this has been replaced by the weaker
conclusion given in (6.1.7). (If it were not for this weakening of the
conclusion of the corresponding (11.1.12), then (6.1.7) would not have
been included, since it follows from (A.3.1) ⇔ (A.3.5) that (6.1.6) ⇔
(6.1.7).)

(6.1) THEOREM. The following statements are equivalent:

(6.1.1) The H-Conjecture (3.3.6) holds: an H-local domain is
catenary.

(6.1.2) If R is an H-local domain, then R' is taut.

(6.1.3) If R is an H-local domain, then there exists a taut
integral extension domain of R .

(6.1.4) If R is an H-local domain, then R satisfies the

o.h.c.c.

(6.1.5) If R is an H-local domain such that R' is level, then R satisfies the s.c.c.

(6.1.6) If R is an H-local domain, then R is an H_2-domain.

(6.1.7) If R is an H-local domain, then R/p is an H-domain, for all height one p ∈ Spec R .

(6.1.8) If R is an H-local domain and R^S is a special extension of R , then $R^S_{M_i}$ is catenary (i = 1,2) .

(6.1.9) If R is an H-local domain, then R^H is taut.

(6.1.10) If R is an H-local domain, then R^* is an H-ring.

(6.1.11) If R is an H-local domain, then $D = R[X]_{(M,X)}$ is an H_2-domain.

(6.1.12) If (R,M) is an H-local domain, then, for all b,c in M such that height (b,c)R = 2 , $B = R[c/b]_{(M,c/b)}$ is an H-domain.

Proof. It is known [McR-2, Propositions 12 and 7] that an integral extension domain of a local domain R is taut if, and only if, R is catenary, so (6.1.1) ⇒ (6.1.2) and (6.1.3) → (6.1.1), and it is clear that (6.1.2) ⇒ (6.1.3).

Since (6.1.1) implies that the Catenary Chain Conjecture holds, by [R-6, (4.5)], (6.1.1) ⇒ (6.1.4). (The reason for referring to [R-6] rather than to (3.3.6) ⇒ (3.3.8) is that in proving (3.3.6) ⇒ (3.3.7) ⇒ (3.3.8), use was made of (6.1.1) ⇒ (6.1.4).)

(6.1.4) implies that if R is an H-local domain and R' is level, then R' satisfies the s.s.c., by (1.3.3), so (6.1.4) ⇒ (6.1.5), by (1.3.4).

Assume (6.1.5) holds and let R be an H-local domain. If R' is level, then (6.1.5) ⇒ (6.1.1), by (1.3.5), so assume that R' is not level. Then altitude R > 1 and there exists a height one maximal ideal in R' , by (3.2), so, with b , S = R' , and B as in (3.1),

$B = R[b,1/b]$ is an H-local domain and B is catenary if and only if R is catenary, by (3.1.2), (3.1.4), and (3.1.6). Also, $B' = R'[1/b]$ is level, by (3.1.1) and (3.2), so (6.1.5) implies that B is catenary, by (1.3.5), hence R is catenary, and so (6.1.1) holds.

It is clear that (6.1.1) $\Rightarrow$ (6.1.6), and (6.1.6) $\Leftrightarrow$ (6.1.7), by (A.3.1) $\Leftrightarrow$ (A.3.5).

Assume (6.1.6) holds and let R be an H-local domain, so R is an H_2-domain. Then, for all height one $p \in \mathrm{Spec}\ R$, altitude R/p = altitude $R - 1$ and R/p is an H-domain, by (6.1.7), so R/p is an H_2-domain, by (6.1.6). Therefore R is an H_3-domain, by (A.3.1) $\Leftrightarrow$ (A.3.5). Repetitions of this show that R is taut-level, so R is catenary, by (A.9.5) $\Rightarrow$ (A.9.1), hence (6.1.1) holds.

(6.1.4) $\Rightarrow$ (6.1.8), by (A.10.1) $\Rightarrow$ (A.10.6) and (1.2.1), and if (6.1.8) holds, then R^s is catenary and height $M_i \in \{1$, altitude $R\}$, by (1.2.2) and (3.2), so R^s is taut, and so (6.1.8) $\Rightarrow$ (6.1.3).

(6.1.4) $\Leftrightarrow$ (6.1.9), by (A.10.1) $\Leftrightarrow$ (A.10.12).

(6.1.4) $\Rightarrow$ (6.1.10), by (A.10.1) $\Rightarrow$ (A.10.15), and (6.1.10) $\Rightarrow$ (6.1.4), by (A.10.15) $\Rightarrow$ (A.10.1) (since R^* is an H-ring only if R^* is taut (since complete local rings are catenary)).

(6.1.4) $\Rightarrow$ (6.1.11), by (A.10.1) $\Rightarrow$ (A.10.19), and (6.1.11) $\Rightarrow$ (6.1.12), by (A.3.1) $\Rightarrow$ (A.3.5).

Finally, assume (6.1.12) holds, let R be an H-local domain, and let P be a height two prime ideal in R. Let b,c in P such that height $(b,c)R = 2$, let $A = R[c/b]$, and let $N = (M,c/b)A$. Then, by [R-9, Lemma 2.7], $P^* = PA$ is a height one prime ideal, $P^* \subseteq N$, and $A/P^* \cong (R/P)[X]$, so height N/P^* = depth $P + 1$. Also, height N = altitude R, by [R-9, Lemma 2.7], so height N/P^* = altitude $R - 1$, by (6.1.12), hence depth P = altitude $R - 2$, so (6.1.12) $\Rightarrow$ (6.1.6), q.e.d.

A different proof that (6.1.1) $\Leftrightarrow$ (6.1.6) was given in [P, (8.7)].

For (6.2), we briefly recall one definition. Namely, if I is
an ideal in a Noetherian ring A , then the <u>Rees</u> <u>ring</u> $\mathcal{R}(A,I)$ <u>of</u> A
<u>with</u> <u>respect</u> <u>to</u> I is the graded subring $\mathcal{R}(A,I) = A[tI,u]$ of $A[t,u]$,
where t is an indeterminate and $u = 1/t$.

(6.2) REMARK. Two additional equivalences of the H-Conjecture
should be mentioned, namely:

(6.2.1) If (R,M) is a C_i-local domain, then $D = R[X]_{(M,X)}$ is
a C_{i+1}-domain; and,

(6.2.2) If R is a C_o-local domain, then, for all b,c in M ,
$\mathcal{R}_m$ is catenary,where $\mathcal{R} = \mathcal{R}(R,(b,c)R)$ is the Rees ring of R with
respect to (b,c)R and m is the maximal homogeneous ideal in $\mathcal{R}$.

<u>Proof</u>. The equivalence of the H-Conjecture and (6.2.1) was given
in [R-9, (3.23)].

The proof of the equivalence of the H-Conjecture and (6.2.2) will
be omitted, since it involves a number of the special properties of
Rees rings.

It should be noted from (4.2.1) and (e) in the introduction to
Chapter 11 that the Chain Conjecture and the Catenary Chain Conjecture
can be characterized in terms of certain DVR's that dominate $R'_{M'}$.
A similar characterization of the H-Conjecture holds, and, in fact,
follows from (6.1.1) $\Leftrightarrow$ (6.1.4) and (A.10.1) $\Leftrightarrow$ (A.10.21).

It was noted in (3.13) that the semi-local H-Conjecture, (3.9.2),
is equivalent to "(3.9.3) and (3.3.6)." Because of this, it is important to know under what circumstances (3.9.3) holds. Now it is shown
in (14.1) and (14.3) that certain of the chain conjectures hold for
local domains of the form $R[X]_{(M,X)}$ (where (R,M) is a local domain)
and for semi-local domains of the form $R[X]_S$ (where R is a semi-
local domain and $S = R[X] - \cup\{(M,X) ; M$ is a maximal ideal in $R\})$.
Thus it might be thought that (3.9.3) should be easier to verify for

$R[X]_S$ - but we show in (6.3) that it is no easier for this case than it is for the general case.

(6.3) REMARK. (3.9.3) is equivalent to (*): if R is a semi-local domain such that $D = R[X]_S$ (with S as above) is an H-semi-local domain, then D_N is an H-domain, for each maximal ideal N in D .

Proof. It is clear that (3.9.3) $\Rightarrow$ (*), so assume that (*) holds and let R be an H-semi-local domain. If there are no height one maximal ideals in R' , then R' is an H-domain, by (3.2), and so D is an H-semi-local domain, by [Hou-2, Theorem 1.9]. Therefore, by (*), for each maximal ideal M in R , $D_{(M,X)D} = R_M[X]_{(MR_M,X)}$ is an H-local domain, so R_M is an H-domain, by (A.3.17) $\Rightarrow$ (A.3.1), and so (3.9.3) holds. Also, if $a =$ altitude $R = 1$, then (3.9.3) holds, so assume that $a > 1$ and that there exists a height one maximal ideal in R' . Let b and B be as in (3.1) (with $S = R'$) , so B is an H-semi-local domain and there are no height one maximal ideals in B' , by (3.1.1) and (3.1.4). Therefore, by what has already been shown, B_P is an H-domain, for each maximal ideal P in B . Let M be a maximal ideal in R . Then, by the choice of b , $(M,1-b)R[b]$ is a maximal ideal in $R[b]$ that lies over M and $(M,1-b)B$ is the only maximal ideal in B that lies over M . Now $R_M \subseteq B_{(M,1-b)B} = $ (say) L , so $B \subseteq R_M[b,1/b] \subseteq L$, and so $L = R_M[b,1/b]$ (since $R_M[b,1/b]$ is local, by the choice of b) . Also, $(M,1-b)B$ is a maximal ideal, so $L = B_{(M,1-b)B}$ is an H-domain, as already noted. Therefore, since there are no height one maximal ideals in L' (since there are none in B') , L' is an H-domain, by (3.2), and so R_M is an H-domain, by (A.3.13) $\Rightarrow$ (A.3.1), q.e.d.

CHAPTER 7

THE DESCENDED GB-CONJECTURE
AND THE GB-CONJECTURE

The Descended GB-Conjecture is new in these notes, but questions
closely related to it were asked in 1976, in [R-10, (3.15) and (3.16)],
[R-12, (4.9)], and [R-19, (7.1.3)]. In [R-10, (3.16)(4)], it was noted
that the Chain Conjecture implies this conjecture. Also, the characterization of this conjecture given in (7.2) was proved in [R-16, (4.3)].

In the comment following (7.1.3) in [R-19], it was pointed out
that with R a local domain as in [N-6, Example 2, pp. 203-205], R
is not a GB-domain, but R' is. (See (B.4.5)). Another such example
was mentioned in (2.9). In both of these cases, R' is not quasi-local
(and is not even level), so some condition (such as in ((3.6.2) or
(3.6.3)) must be placed on R' when considering a conjecture of this
type.

The history of the "non-Noetherian" GB-Conjecture was given in
(2.1). The "Noetherian" GB-Conjecture (given in (3.6.4)) was essentially asked in 1972, by I. Kaplansky in the introduction to [K], and
I somewhat more specifically asked it in 1976, in the comment between
(2.1) and (2.2) in [R-16]. In (4.2) of this latter paper this conjecture was essentially characterized as in (7.4.3).

Finally, both of these conjectures hold for local domains that
satisfy the s.c.c., by [R-16, (3.10) and (3.1)], but I. Kaplansky's
example in [K] shows that there exist integrally closed quasi-local domains for which neither of these conjectures hold.

(7.1) gives two characterizations of the Descended Chain Conjecture. For this result and for (7.4), the reader should refer to (A.6)
for some equivalences of the GB condition.

(7.1) THEOREM. The following statements are equivalent:

(7.1.1) The Descended GB-Conjecture (3.6.3) holds: if R is a local domain such that R' is quasi-local, then R is a GB-domain.

(7.1.2) If R is a local domain such that R' is quasi-local, and if S = R[c] is a principal integral extension domain of R such that $c^2 + rc \in R$, for some $r \in R$, then adjacent prime ideals in S lie over adjacent prime ideals in R .

(7.1.3) With R as in (7.1.2), R' is a GB-domain and adjacent prime ideals in R' lie over adjacent prime ideals in R .

Proof. It is clear that (7.1.1) ⇒ (7.1.2), and the proof that (7.1.2) ⇒ (7.1.1) is similar to (and easier than) the proof that (7.4.4) ⇒ (7.4.1), so it will be omitted.

If (7.1.1) holds, then R' is a GB-domain, by (A.6.1) ⇒ (A.6.2), so (7.1.3) holds by the definition of a GB-domain. And it readily follows from the Going Up Theorem that (7.1.3) ⇒ (7.1.1), q.e.d.

(7.2) REMARK. In [R-16, (4.3)] it was shown that the Descended GB-Conjecture is equivalent to: if R is a local domain such that R' is quasi-local, then, for all prime ideals $P \subset Q$ in R such that height Q/P > 1 , the Upper Conjecture (3.8.5) holds for $L = R_Q/PR_Q$ and there are no height one maximal ideals in L' .

We next show that this conjecture could have been stated for a more general case.

(7.3) PROPOSITION. The Descended GB-Conjecture (7.1.1) is equivalent to (*): if A is a Noetherian domain such that there exists a one-to-one correspondence between the maximal ideals in A and A' , then A is a GB-domain.

Proof. It is clear that (*) ⇒ (7.1.1), so assume that (7.1.1) holds and let A be a Noetherian domain such that there exists a one-to-one correspondence between the maximal ideals in A and A' . Then, to prove that A is a GB-domain, it suffices to prove that, for each

maximal ideal M in A , A_M is a GB-domain, by (A.6.1) $\Leftrightarrow$ (A.6.5),
and this follows immediately from the hypothesis, q.e.d.

The last theorem in this chapter gives three characterizations of
the GB-Conjecture.

(7.4) THEOREM. The following statements are equivalent:

(7.4.1) The GB-Conjecture (3.6.4) holds: the integral closure
of a Noetherian domain is a GB-domain.

(7.4.2) If R is a local domain such that R' is level, then
R' is a GB-domain.

(7.4.3) If R is a local domain such that R' is quasi-local,
and if S = R'[c] is a free integral extension domain of R' , then
adjacent prime ideals in S lie over adjacent prime ideals in R' .

(7.4.4) With R and S as in (7.4.3), the conclusion of (7.4.3)
holds when c is quadratic over R' .

Proof. It is clear that (7.4.1) $\Rightarrow$ (7.4.2) $\Rightarrow$ (7.4.3) $\Rightarrow$ (7.4.4),
so assume (7.4.4) holds and let R be a Noetherian domain. Then, to
prove that R' is a GB-domain it is sufficient to prove that $R'_{M'}$
is a GB-domain, for each maximal ideal M' in R' , by (A.6.1) $\Rightarrow$
(A.6.5), so it may be assumed that R is local and R' is quasi-local
(as in the proof of (3.6.3) $\Rightarrow$ (3.6.4)). Therefore, suppose there exist
adjacent prime ideals $P_0 \subset Q_0$ in some integral extension domain of
R' such that height q/p > 1 , where $q = Q_0 \cap R'$ and $p = P_0 \cap R'$.
Then, with $L = (R'/p)_{q/p}$, there exists a height one maximal ideal in
L', by [R-4, Lemma 2.9] (since L lies between a Noetherian domain and
its integral closure and since there is a height one maximal ideal in
some integral extension domain of L) . For these same reasons, it follows from [R-2, Proposition 3.5 and Remark 3.2] that there exist $b \in L'$
and $s \in L$ such that $b^2 - sb \in L$ and L[b] has a height one maximal
ideal. Let K^0 be the kernel of the natural homomorphism of L[X]

onto L[b] and let K be the pre-image of K^o in $R'_q[X]$. Then
$pR'_q[X] \subset K$, $K \cap R'_q = pR'_q$, and there exists a maximal ideal N in
$R'_q[X]$ such that $K \subset N$ and height N/K = 1 . Also, K contains a
monic quadratic polynomial f , so let p^* be a height one prime
ideal in $R'_q[X]$ such that $f \in p^* \subset N$. Let $A = R'_q[X]/p^*$, P =
K/p^* , and $Q = N/p^*$. Then A is an integral extension domain of
R'_q , $P \subset Q$, height Q/P = 1 , $P \cap R'_q = pR'_q$, and $Q \cap R'_q = qR'_q$ (since
Q is maximal and A is integral over R'_q) . Now $pR'_q \subset qR'_q$ are
not adjacent prime ideals, so $R'_q \subset A$. Therefore $p^* = (f)$ (by the
division algorithm, since f is quadratic and monic and p^* contains
no linear polynomials (since R'_q is integrally closed)). Therefore
A is a free R'_q-algebra, so it follows that there exists a free quadratic integral extension domain of R' that has adjacent prime ideals
that contract in R' to non-adjacent prime ideals, and this contradicts (7.4.4). Therefore (7.4.4) $\Rightarrow$ (7.4.1), q.e.d.

(7.5) REMARK. It readily follows from the definition that the
following statements are equivalent for an integral domain A : A is
a GB-domain; each saturated chain of prime ideals in each integral extension domain of A contracts in A to a saturated chain of prime
ideals; and, each maximal chain of prime ideals in each integral extension domain of A contracts in A to a maximal chain of prime ideals.
Using this together with the rings in (7.1) and (7.4), the reader can
readily give some additional equivalences of the Descended GB-Conjecture and the GB-Conjecture.

CHAPTER 8

THE STRONG AVOIDANCE CONJECTURE
AND THE AVOIDANCE CONJECTURE

The Strong Avoidance Conjecture is new in these notes, and it results from combining the Avoidance Conjecture and the semi-local Depth Conjecture. (See $(3.8.2) \Rightarrow (3.8.3)$ and $(3.14.1)$.)

The Avoidance Conjecture arose by imposing an additional condition on the following familiar result in commutative algebra [ZS-1, Lemma, p. 240]: if $P_1 \subset P_2 \subset \cdots \subset P_n$ is a chain of prime ideals in a Noetherian ring A and $N_1, \ldots, N_h$ are prime ideals in A such that $P_n \not\subseteq \cup N_i$, then there exists a chain of prime ideals $P_1 \subset P_2' \subset \cdots \subset P_n' = P_n$ such that $P_j' \not\subseteq \cup N_i$, for $j = 2, \ldots, n$.

This conjecture was first stated by S. McAdam in 1974, in [Mc-1, Question 1, p. 728], and in the more general form of (3.8.3) in 1975, in [HMc, p. 752]. As mentioned after the proof of (3.8), it was shown in [HMc, Proposition 3.8] that this conjecture implies the Upper Conjecture, and it was noted in [Mc-1, p. 728] that this conjecture implies the Taut-Level Conjecture. Also, in [Mc-4, Theorem 3], it was recently shown that this conjecture holds for non-extended prime ideals Q of **little-height** two in $R[X]$ (that is, $Q \in \operatorname{Spec} R[X]$ is such that $(Q \cap R)R[X] \neq Q$ and there exists a saturated chain of prime ideals in $R[X]$ of the form $(0) \subset p \subset Q$). (Some results that are closely related to this last result are given in [Mc-2].)

Finally, by [ZS-1, Lemma, p. 240], the Strong Avoidance Conjecture holds for catenary local domains and the Avoidance Conjecture holds for catenary Noetherian rings. On the other hand, in (14.7) it is shown that there exists a quasi-local domain for which neither of these conjectures hold.

(8.1) gives one equigalence of the Strong Avoidance Conjecture. (Two additional equivalences follow from (8.2).)

(8.1) THEOREM. The following statements are equivalent:

(8.1.1) The Strong Avoidance Conjecture (3.8.2) holds: if $P \subset$ $Q \subset N$ is a saturated chain of prime ideals in a semi-local ring R and if $N_1,\ldots,N_h$ are prime ideals in R such that $N \not\subseteq \cup N_i$, then there exists $q \in$ Spec R such that $P \subset q \subset N$, $q \not\subseteq \cup N_i$, height q = height P + 1 , and depth q = depth N + 1 (so $P \subset q \subset N$ is saturated).

(8.1.2) If (0) $\subset Q \subset N$ is a saturated chain of prime ideals in a semi-local domain R and if $N_1,\ldots,N_h$ are prime ideals in R such that $N \not\subseteq \cup N_i$, then there exists a height one $q \in$ Spec R such that $q \subset N$, $q \not\subseteq \cup N_i$, and depth q = depth N + 1 .

Proof. It is clear that (8.1.1) $\Rightarrow$ (8.1.2), so assume that (8.1.2) holds and let the notation be as in (8.1.1). Then it follows from (8.1.2) and (5.2) (on passing to R/P) that there exist infinitely many $q \in$ Spec R such that $P \subset q \subset N$ is saturated, $q \not\subseteq \cup N_i$, and depth q = depth N + 1 . Therefore, since only finitely many of these q are such that height q > height P + 1 , by [Mc-1, Theorem 1], (8.1.1) holds, q.e.d.

(8.2) REMARK. By (5.2), the "there exists" in (8.1.1) and in (8.1.2) can be replaced by "there are infinitely many."

(8.3) gives one equivalence of the Avoidance Conjecture, and (8.4) provides two additional equivalences.

(8.3) THEOREM. The following statements are equivalent:

(8.3.1) The Avoidance Conjecture (3.8.3) holds: if $P \subset Q \subset N$ is a saturated chain of prime ideals in a Noetherian ring A and if $N_1,\ldots,N_h$ are prime ideals in A such that $N \not\subseteq \cup N_i$, then there exists $q \in$ Spec A such that $P \subset q \subset N$ is saturated and $q \not\subseteq \cup N_i$.

(8.3.2) If (0) $\subset p \subset N$ is a maximal chain of prime ideals in a semi-local domain R , then there exists $q \in$ Spec R such that

$(0) \subset q \subset N$ is saturated and N is the only maximal ideal in R that contains q .

Proof. It is clear that $(8.3.1) \Rightarrow (8.3.2)$, so assume $(8.3.2)$ holds and let A , $P \subset Q \subset N$, and $N_1, \ldots, N_h$ be as in $(8.3.1)$. Let $R = A_S/PA_S$, where $S = A - (N \cup N_1 \cup \cdots \cup N_h)$. Then it follows from $(8.3.2)$ that there exists $q \in \text{Spec } A$ such that $P \subset q \subset N$ is saturated and $q \not\subseteq \cup N_i$, so $(8.3.1)$ holds, q.e.d.

(8.4) REMARK. The "there exists" in $(8.3.1)$ and in $(8.3.2)$ can be replaced by "there are infinitely many."

Proof. The proof of [Mc-1, Proposition 2] shows that with $P \subset Q \subset N$, $N_1, \ldots, N_h$, and A as in $(8.3.1)$ and with: $S = A - (N \cup N_1 \cup \cdots \cup N_h)$; $R = A_S/PA_S$; and, $M = NA_S/PA_S$; there are infinitely many $q \in \text{Spec } R$ such that $(0) \subset q \subset M$ is a mcpil 2 and M is the only maximal ideal in R that contains q , if there is one such q . The conclusion readily follows from this, q.e.d.

Concerning $(8.1.2)$ and $(8.3.2)$, it is known [R-4, Proposition 2.2] that if $(0) \subset P \subset Q$ is a saturated chain of prime ideals in a local domain R , then $I = \{p \in \text{Spec } R ; (0) \subset p \subset Q \text{ is saturated}\}$ is an infinite set. However, it is shown in [HMc, Example 3.2] that Q may properly contain $\cup\{p ; p \in I\}$. (See $(B.3.9)$.)

CHAPTER 9

THE UPPER CONJECTURE

(A nonextended prime ideal Q in R[X] is said to be an upper
to Q∩R , and the name of this conjecture derives from this.) M.
Nagata's examples [N-6, Example 2, pp. 203-205] show that if (R,M)
is a local domain, then there may exist a mcpil 2 in $D = R[X]_{(M,X)}$
when there does not exist a mcpil 1 in R (that is, when altitude
R > 1) , but for a mcpil n + 1 > 2 in D , there exists a mcpil n
in R (in his examples). Since this also holds for all other local
domains for which this has been tested, the Upper Conjecture arose from
asking the natural question.

This conjecture was first stated by S. McAdam in 1974, in [Mc-1,
Question 3, p. 728], and it was stated in [HMc, p. 750] as: if (R,M)
is a local domain and Q is an upper to M in R[X] , then {m ; there
exists a mcpil m in $R[X]_Q$} ⊆ {2} ∪ {n+1 ; there exists a mcpil n
in R] . Now it is known [R-10, (5.5.6)] that there exists a mcpil m
in $R[X]_Q$ if, and only if, there exists a mcpil m in $R[X]_{(M,X)}$,
so (3.8.5) is equivalent to the statement of this conjecture as given
in [HMc, p. 750]. This conjecture was stated (essentially) as in (3.8.5)
in [RMc, (2.21)] and in [R-12, (4.10.3)], and (9.2) contains a number
of equivalences of this conjecture that were given in these two papers.

Finally, in (14.1) it is shown that this conjecture holds for Henselian local domains and local domains of the form $L[X]_{(M,X)}$, while
in (14.5) it is shown that there exists a quasi-local domain for which
this conjecture does not hold.

The characterizations of the Upper Conjecture in (9.1) and (9.2)
involve the existence of a mcpil n in a local domain. The reader is
referred to (A.5) for some equivalences of this condition.

(9.1) THEOREM. The following statements are equivalent:

(9.1.1) The Upper Conjecture (3.8.5) holds: if (R,M) is a local domain and if there exists a mcpil n+1 $\underline{in}$ $D = R[X]_{(M,X)}$, $\underline{then}$ either there exists a mcpil n $\underline{in}$ R $\underline{or}$ n = 1 .

(9.1.2) $\underline{If}$ R $\underline{is}$ $\underline{a}$ local domain and there exists a mcpil n > 1 $\underline{in}$ R' , $\underline{then}$ there exists $\underline{a}$ mcpil n $\underline{in}$ R .

(9.1.3) $\underline{If}$ R $\underline{is}$ $\underline{a}$ local domain and there exists a mcpil n > 1 $\underline{in}$ $\underline{a}$ principal integral extension domain R[c] $\underline{of}$ R such that R[c] $\subseteq$ R' , $\underline{then}$ there exists $\underline{a}$ mcpil n $\underline{in}$ R .

(9.1.4) $\underline{If}$ R $\underline{is}$ $\underline{a}$ local domain and there exists $\underline{a}$ mcpil n > 1 $\underline{in}$ $\underline{a}$ special extension R^s $\underline{of}$ R , $\underline{then}$ there exists $\underline{a}$ mcpil n $\underline{in}$ R .

(9.1.5) $\underline{If}$ (R,M) $\underline{is}$ $\underline{a}$ local domain and there exists $\underline{a}$ mcpil n $\underline{in}$ $(R_k)_Q$, $\underline{where}$ $R_k = R[X_1,\ldots,X_k]$ (k > 0) $\underline{and}$ Q $\in$ Spec R_k $\underline{is}$ $\underline{such}$ $\underline{that}$ $MR_k \subset Q$, $\underline{then}$ $\underline{either}$ there exists $\underline{a}$ mcpil n - k + depth Q $\underline{in}$ R $\underline{or}$ n - k + depth Q = 1 .

(9.1.6) $\underline{If}$ (R,M) $\underline{is}$ $\underline{a}$ local domain and there exists $\underline{a}$ mcpil n+1 > 2 $\underline{in}$ $R[X]_{(M,X)}$, $\underline{then}$ there exists $\underline{a}$ mcpil n $\underline{in}$ R .

Proof. (9.1.1) $\Rightarrow$ (9.1.2), by (9.2.1) $\Leftrightarrow$ (9.2.3), (9.1.2) $\Rightarrow$ (9.1.3), by the Going Up Theorem, and it is clear that (9.1.3) $\Rightarrow$ (9.1.4).

(9.1.4) $\Rightarrow$ (9.1.1): By (9.2.3) $\Rightarrow$ (9.2.1), it suffices to prove if R is a local domain and there exists a mcpil n > 1 in some integral extension domain of R , then there exists a mcpil n in R . For this, [HMc, Theorem 1.10] says there exists a principal integral extension domain A = R[c] of R that has a mcpil n . If A is local, then there exists a mcpil n in R , by (A.5.2) $\Rightarrow$ (A.5.1), so assume A is not local. Let (0) $\subset$ P_1 $\subset$ $\cdots$ $\subset$ P_n be a mcpil n in A , let d $\in$ P_n such that 1 - d is in all other maximal ideals in A , and let C = R[d] . Then A_{P_n} is integral over $C_{P_n \cap C}$, so there exists a mcpil n in $C_{P_n \cap C}$, by (A.5.2) $\Rightarrow$ (A.5.1), hence there exists a mcpil n in C . Let L = R + J , where J is the Jacobson radical of C . Then L is local and it is readily seen that C is a special extension of L , so there exists a mcpil n in L , by hypothesis, and so there exists a mcpil n in R , by (A.5.2) $\Rightarrow$ (A.5.1).

With the notation of (9.1.5), there exists a mcpil n in $(R_k)_Q$ if and only if there exists a minimal prime ideal z in R^* such that depth z = n - k + depth Q , by (1.5.2) $\Leftrightarrow$ (1.5.3) , so (9.1.5) $\Leftrightarrow$ (9.1.1), by (9.2.1) $\Leftrightarrow$ (9.2.2).

Finally, it is clear that the Upper Conjecture implies (9.1.6), so assume that (9.1.6) holds, let (R,M) be a local domain, and assume that there exists a mcpil n+1 in $D = R[X]_{(M,X)}$. If n+1 > 2 , then there exists a mcpil n in R as desired, by (9.1.6). If n+1 $\leq$ 2 , then either n = 0 or n = 1 . If n = 1 , then (9.1.1) holds, so assume that n = 0 . Then it follows that R is a field, so there exists a mcpil n = 0 in R , and so (9.1.1) holds, q.e.d.

In (9.2) we list four other known equivalences of this conjecture.

(9.2) REMARK. The following statements are equivalent:

(9.2.1) The Upper Conjecture (9.1.1) holds.

(9.2.2) If R is a local domain and there exists a depth n > 1 minimal prime ideal in R^* , then there exists a mcpil n in R .

(9.2.3) If R is a local domain and there exists an integral extension domain of R that has a mcpil n > 1 , then there exists a mcpil n in R .

(9.2.4) $\mathcal{S} = \mathcal{J}$, where $\mathcal{S}$ = {R ; R is a local domain such that there exists a mcpil n in some integral extension domain of R (if and) only if there exists a mcpil n in R} and $\mathcal{J}$ = {R ; R is a local domain and either altitude R = 1 or there are no height one maximal ideals in R'} .

Proof. (9.2.1) $\Leftrightarrow$ (9.2.2) $\Leftrightarrow$ (9.2.3), by [RMc, (2.22)], and (9.2.1) $\Leftrightarrow$ (9.2.4), by [R-12, (4.10.3)], q.e.d.

(7.5) shows that if the GB-Conjecture holds and R is an inte-

grally closed local domain, then R satisfies the condition in (9.2.3),
so the GB-Conjecture implies R satisfies the Upper Conjecture, by
(9.2.1) ⇔ (9.2.3).

The set $\mathcal{S}$ of (9.2.4) is of some interest, since every GB-local
domain is in it. Some questions concerning $\mathcal{S}$ are mentioned in
(15.4.5), and in (B.3.5) it is shown that $\mathcal{S}$ is not closed under factorization nor under localization - that is, $R \in \mathcal{S}$ and $P \in$ Spec R
does not imply that R/P or R_p are in $\mathcal{S}$.

The final result in this chapter shows that this conjecture is
equivalent to its semi-local version.

(9.3) THEOREM. The following statements are equivalent:

(9.3.1) The Upper Conjecture (9.1.1) holds.

(9.3.2) If R is a semi-local domain and there exists a mcpil
n+1 in $R[X]_S$, where $S = R[X] - \cup\{(M,X)$; M is a maximal ideal in
R} , then either there exists a mcpil n in R or n = 1 .

(9.3.3) If R is a semi-local domain and there exists a mcpil
n+1 > 2 in $R\langle X \rangle$, where $R\langle X \rangle = R[X]_{S'}$ with $S' = R[X] - \cup\{N$; N
and N∩R are maximal ideals} , then there exists a mcpil n in R .

Proof. Assume (9.3.1) holds, let R be a semi-local domain, and
assume there exists a mcpil n+1 > 2 in $R\langle X \rangle$. Then there exists a
maximal ideal N in $R\langle X \rangle$ such that there exists a mcpil n+1 in
$R\langle X \rangle_N$. Let $M = N \cap R$ and $N^* = NR\langle X \rangle_N \cap R_M[X]$. Then M is a maximal
ideal in R , by [Hou-2, Preliminaries], and N^* is an upper to MR_M ,
so there exists a mcpil n+1 > 2 in $R_M[X]_{(MR_M, X)}$, by [R-10, (5.5.6)].
Therefore there exists a mcpil n in R_M , by (9.3.1), so there exists
a mcpil n in R , and so (9.3.3) holds.

Assume (9.3.3) holds, let $D = R[X]_S$ be as in (9.3.2), and assume
there exists a mcpil n+1 in D . Then, since D is a quotient ring
of $R\langle X \rangle$ and the maximal ideals in D lie over maximal ideals in

$R\langle X\rangle$, there exists a mcpil n+1 in $R\langle X\rangle$. Therefore, if n+1 > 2 , then there exists a mcpil n in R , by (9.3.3). If n+1 ≦ 2 , then clearly n ∈ {0,1} . If n = 1 , then (9.3.2) holds, so assume that n = 0 . Then it follows that R is a field, so there exists a mcpil n = 0 in R , and so (9.3.2) holds.

Finally, it is clear that (9.3.2) ⟹ (9.3.1), q.e.d.

CHAPTER 10
THE TAUT-LEVEL CONJECTURE

It is known [McR-2, Proposition 12] that an integral extension domain of a catenary local domain R is taut, and is taut-level, if R' is level (by [McR-2, Proposition 12] and (3.2)). Therefore, since the Catenary Chain Conjecture holds if every integral extension domain of R is catenary, by [R-6, (4.3)], the Taut-Level Conjecture arose from generalizing the natural question to ask about integral extension domains of a catenary local domain.

I stated this conjecture in 1971, in [R-4, Remark 3.14(iii)], and it was more explicitly stated in [Mc-1, Question 2, p. 728], in Remark (i) preceding Proposition 9 in [McR-2], and in [R-14, (2.14)]. In these last two references, it was noted without proof that (10.1.1) ⇔ (10.1.7) (in [McR-2]), that (10.1.2) ⇔ (10.1.6) (in [R-14]), and that these equivalent conditions imply the Catenary Chain Conjecture. Also, it was noted in [R-4, Remark 3.14(iii)] (together with [McR-2, Propositions 12 and 13]) that this conjecture implies the Normal Chain Conjecture. Moreover, a remark relating this conjecture to Conjecture (K) (see (13.1)) was given in [McR-1, (2.10)], and in [R-21, (3.9)] it was shown that this conjecture holds, if, for each taut semi-local domain R and for all maximal ideals N in $I = \cap\{V \; ; \; (V,N) \in \mathcal{U}\}$ (with $\mathcal{U}$ as in (4.2.2) (generalized to semi-local domains)), I_N is an H-domain.

Finally, in (14.3) it is shown that the Taut-Level Conjecture holds for Henselian semi-local rings and certain localizations of $R[X]$. But, on the other hand, an example of a taut-level semi-local ring (not a domain) which is not catenary was given preceding Corollary 8 in [McR-2], and the existence of additional examples showing the conjecture must be restricted to semi-local domains has already been noted in (2.2) and (2.8).

Seven characterizations of this conjecture are given in (10.1).

76

The reader should refer to (A.7) - (A.11) for some equivalences of the various conditions mentioned in (10.1).

(10.1) THEOREM. The following statements are equivalent:

(10.1.1) The Taut-Level Conjecture (3.9.5) holds: if R is a taut-level semi-local domain, then R satisfies the f.c.c.

(10.1.2) If R is a taut semi-local domain, then R is catenary.

(10.1.3) If R is a taut semi-local domain, then R' satisfies the c.c.

(10.1.4) If R is a taut semi-local domain, then R satisfies the o.h.c.c.

(10.1.5) If R is a taut-semi-local domain and R' is level, then R satisfies the s.c.c.

(10.1.6) If R is a taut semi-local domain, then R_M is taut, for all maximal ideals M in R .

(10.1.7) If R is a taut-level semi-local domain, then R/P is taut-level, for all P $\in$ Spec R .

(10.1.8) If R is a taut semi-local domain such that R' is level, then D = $R[X]_S$ is taut, where S is the complement in R[X] of the union of the ideals (M,X)R[X] with M maximal in R .

Proof. Assume (10.1.1) holds and let R be a taut semi-local domain. If R' is level, then R is level, so R is catenary, by (10.1.1), and so (10.1.2) holds. Therefore, assume R' is not level, so altitude R > 1 and there exists a height one maximal ideal in R' , by (3.2). Let b , S = R' , and B = R[b,1/b] be as in (3.1), so B is a taut-level semi-local domain, by (3.1.1) and (3.1.5). Therefore B satisfies the f.c.c., by (10.1.1), so R is catenary, by (3.1.6), and so (10.1.2) holds.

(10.1.2) $\Rightarrow$ (10.1.4): Assume (10.1.2) holds and let R be a taut semi-local domain. If R' is level, then R is level, so R is cate-

nary and level, by (10.1.2), and so R satisfies the f.c.c., by (1.2.3).
Also, each finite integral extension domain A of R is taut (by
(A.8.1) ⇔ (A.8.3)) and A' is level (by [N-6, (10.14)], since R' is
level), so A is catenary and level, by (10.1.2), and so A satisfies
the f.c.c. Therefore R satisfies the s.c.c., by (A.11.1) ⇔ (A.11.8),
so R' satisfies the s.c.c., by (A.11.1) ⇒ (A.11.7), and so (10.1.4)
holds. Therefore assume R' is not level, so altitude R > 1 and
there exists a height one maximal ideal in R' , by (3.2). Let b ,
S = R' , and B be as in (3.1), so B = R[b,1/b] is a taut-level semi-
local domain such that B' has no height one maximal ideals, by (3.1.1)
and (3.1.5). Therefore B' is level, by (3.2), so by what has already
been shown, B' satisfies the s.c.c., hence R' satisfies the c.c.,
by (3.1.7), so (10.1.4) holds.

It is clear that (10.1.4) ⇒ (10.1.3), and (10.1.3) implies that if
R is a taut semi-local domain such that R' is level, then R' satisfies the s.c.c., by (1.3.3), so (10.1.3) ⇒ (10.1.5), by (1.3.4).

Assume (10.1.5) holds and let R be a taut-level semi-local domain.
If R' is level, then (10.1.1) holds, by (10.1.5) and (1.3.5). Therefore, assume R' is not level and let b , S = R' , and B be as in
the proof that (10.1.2) ⇒ (10.1.4). Then B is taut-level and B' is
level, hence B satisfies the f.c.c., by (10.1.5) and (1.3.5), and so
R is catenary, by (3.1.6). Therefore R is level and catenary, hence
R satisfies the f.c.c., by (1.2.3), and so (10.1.1) holds.

If (10.1.2) holds and R is as in (10.1.6), then, for each maximal
ideal M in R , R_M satisfies the f.c.c., by (1.2.2), so (10.1.6) holds.

(10.1.6) implies that, for each maximal ideal M in a taut semi-
local domain R , R_M satisfies the f.c.c., by (A.9.5) ⇒ (A.9.1), so
(10.1.6) ⇒ (10.1.2), by (1.2.2).

Assume (10.1.1) holds, let R be a taut-level semi-local domain,
and let P ∈ Spec R . Then R/P satisfies the f.c.c. (by (1.2.1),

since R does), so R/P is taut-level, hence (10.1.7) holds.

Assume (10.1.7) holds, let R be a taut-level semi-local domain, and let $(0) \subset P_1 \subset \cdots \subset P_k$ be a maximal chain of prime ideals in R. Then, by hypothesis, altitude $R - 1 =$ depth $P_1 =$ height P_k/P_1. Also, by induction on altitude R , R/P_1 satisfies the f.c.c., so height $P_k/P_1 = k - 1$. Therefore $k =$ altitude R , so R satisfies the f.c.c., hence (10.1.1) holds.

If (10.1.5) holds and R is a taut semi-local domain such that R' is level, then, for all maximal ideals M in R , height M = altitude R and R_M satisfies the s.c.c., by hypothesis and (1.3.3). So, by (A.11.1) ⇒ (A.11.19), $R[X]_{(M,X)} = R_M[X]_{(MR_M,X)}$ is catenary and altitude $R[X]_{(M,X)} =$ altitude $R + 1$. Therefore, it follows that $R[X]_S$ is taut, and so (10.1.8) holds.

Finally, assume (10.1.8) holds, let R be as in (10.1.5), and let M be a maximal ideal in R . Then height M = height MD = altitude D = 1 = altitude R and $R_M(X)$ $(= R_M[X]_{MR_M[X]}) = D_{MD}$, so $R_M(X)$ satisfies the s.c.c., by (A.8.1) ⇒ (A.8.5). Therefore R_M satisfies the s.c.c., by [RMc , (2.15)], hence R satisfies the s.c.c., by (1.3.3), and so (10.1.5) holds, q.e.d.

The proof of (10.1.5) ⇒ (10.1.8) shows that, in fact, D is taut-level, so it follows that (10.1.8) is equivalent to: if R is a taut semi-local domain such that R' is level, then D is taut-level.

The proof that (10.1.8) ⇒ (10.1.5) given above is the same as that given in [R-14, (4.29.3)].

CHAPTER 11

THE CATENARY CHAIN CONJECTURE

I stated this conjecture in 1971 in the introduction of [R-4].
The original reason for considering this conjecture was that if R is
a catenary local domain, then, for all nonmaximal prime ideals P' in
R' , height P' + depth P' = altitude R' and $R'_{P'}$ satisfies the s.c.c.,
by (A.8.1) $\Rightarrow$ (A.8.3) and [R-4, Corollary 3.12] - so it seems that it
should be possible to prove that R' satisfies the c.c. from this and
the fact that R' is a Krull domain.

One equivalence of this conjecture has already been given in
(3.10.3) and a number of equivalences of this conjecture have previously
appeared in the literature. For example:

(a)   every special extension of a catenary local domain is catenary
[R-6, (4.7)];

(b)   if S is a finite integral extension domain of a catenary
local domain R , then $S_N[1/b]$ satisfies the s.c.c., for all maximal
ideals N in S and for all nonzero b in NS_N [R-11, (2.12.2)];

(c)   if (R,M) is a catenary local domain and R' is level, then,
for all nonzero b in M , $\mathcal{R}_{\mathfrak{m}}$ is catenary, where $\mathcal{R} = \mathcal{R}(R,bR)$ is
the Rees ring of R with respect to bR and $\mathfrak{m}$ is the maximal homo-
geneous ideal in $\mathcal{R}$ [R-9, Question 3.22];

(d)   if R is a catenary local domain, then R^{*} is taut [R-14,
(3.18)]; and,

(e)   if R is a catenary local domain, then, for each maximal
ideal M' in R' , every DVR (V,N) in the quotient field of R such
that $R \subseteq V$, $N \cap R' = M'$, and V is integral over a locality over R
is of the first kind [R-13, (2.19.2)].

(See also [R-6, (4.3)] and [P,(8.8)].) Of these, (a) and (d) are
sharpened in (11.1.8) and (11.1.10), respectively, and (e) follows from
(11.1.1) $\Leftrightarrow$ (11.1.3) and (A.10.1) $\Leftrightarrow$ (A.10.21).

Also, this conjecture holds, if, for all $P \in \operatorname{Spec} R$, R_P is an H-domain, whenever R is an H-local domain, by [R-20, last paragraph], and a remark relating this conjecture to Conjecture (K) (see (13.1)) was given in [McR-1, (2.9)]. And, as noted after the proof of (3.3), it was proved in [R-5] and in [R-6] that the H-Conjecture implies this conjecture, and it was noted without proof in the introduction of [R-4] that this conjecture implies the Normal Chain Conjecture.

It is known [R-3, Theorem 2.21] that if L is a Henselian local ring, a complete local ring, or a local ring of the form $R[X]_{(M,X)}$ with (R,M) a local ring), then L satisfies the f.c.c. if, and only if, L satisfies the s.c.c. (so this conjecture holds for such L) . It is interesting to note that it is shown in (11.1.1) $\Leftrightarrow$ (11.1.4) that one of the characterizations of the Catenary Chain Conjecture is that the f.c.c. and the s.c.c. are equivalent conditions for all local domains R such that R' is level. Eleven other characterizations of this conjecture are given in (11.1). (It should be noted that the statements are arranged to correspond to the order of the statements in the characterizations of the H-Conjecture, (6.1), and this is quite different from the order in which they are proved to be equivalent. Also, the reader is referred to (A.3), (A.4), and (A.9) - (A.11) for some equivalences of the various conditions mentioned in (11.1).)

(11.1) THEOREM. The following statements are equivalent:

(11.1.1) The Catenary Chain Conjecture (3.3.8) holds: the integral closure of a catenary local domain satisfies the c.c.

(11.1.2) If R is a catenary local domain, then R' is catenary.

(11.1.3) If R is a catenary local domain, then R satisfies the o.h.c.c.

(11.1.4) If R is a catenary local domain such that R' is level, then R satisfies the s.c.c.

(11.1.5) If R is a catenary local domain, then R is a C_1-

domain.

(11.1.6) If a ring A is taut (respectively, taut-level) and is a finite integral extension domain of a local domain R , then A is catenary (respectively, satisfies the f.c.c.).

(11.1.7) If R is a catenary local domain, then $R'_{M'}$ is an H-domain, for all maximal ideals M' in R' .

(11.1.8) If R is a catenary local domain and R^s is a special extension of R , then $R^s_{M_i}$ is an H-domain (i = 1,2) .

(11.1.9) If R is a catenary local domain, then R^H is an H-ring.

(11.1.10) If R is a catenary local domain, then R^* is an H-ring.

(11.1.11) If (R,M) is a catenary local domain, then D = $R[X]_{(M,X)}$ is an H_2-domain.

(11.1.12) With R and D as in (11.1.11), for each height one prime ideal p in R , D/pD is an H-domain.

(11.1.13) If R is a catenary local domain, then, for all b,c in M such that height (b,c)R = 2, B = $R[c/b]_{(M,c/b)}$ is an H-domain.

Proof. It is clear that (11.1.1) ⇒ (11.1.3), (11.1.3) ⇒ (11.1.2), by (1.3.5), and (11.1.2) ⇒ (11.1.7), by (1.2.1).

Assume (11.1.7) holds and let $(R^s ; M_1, M_2)$ be a special extension of a catenary local domain R . Fix i , let $N = M_i$, and let p ∈ Spec R^s such that height p = 1 and p ⊆ N . Then to prove that R^s_N is an H-domain, it may be assumed that height N = altitude R > 1, by (3.2). Now R ⊂ R^s ⊆ R' , so there exists a height one p' ∈ Spec R' such that p'∩R^s = p . Then p' ⊂ M' , for some maximal ideal M' in R' , and height M' = altitude R , by (3.2). Therefore height M'/p' = altitude R - 1 , by (11.1.7), hence it follows that height N/p = altitude R - 1 , so R^s_N is an H-domain, and so (11.1.8) holds.

Assume that (11.1.8) holds. Then to prove that (11.1.1) holds, it suffices to prove that if R is a catenary local domain and R^S is a special extension of R , then R^S is catenary, by [R-6, (4.7)]; that is, that R^S_M is catenary, for $i = 1,2$, by (1.2.2). Fix i and let $N = M_i$, so height $N \in \{1$, altitude $R\}$, by (3.2). Therefore to prove that R^S_N is catenary it may clearly be assumed that height $N =$ altitude R , and it suffices to prove that if p is a height one prime ideal in R^S such that $p \subset N$, then height $N/p =$ altitude $R - 1$ and R^S/p is catenary, by (A.9.1) $\Rightarrow$ (A.9.6). For this, height $N/p =$ altitude $R - 1$, by (11.1.8) (since height $N =$ altitude R) . Also, either $R^S/p = R/(p \cap R)$ or R^S/p is a special extension of $R/(p \cap R)$, by [R-6, (4.8)(2)]. If $R^S/p = R/(p \cap R)$, then R^S/p is catenary, since $R/(p \cap R)$ is, by (1.2.1). On the other hand, if R^S/p is a special extension of $R/(p \cap R)$, then, by induction on altitude R , R^S/p is catenary. Therefore it follows that R^S is catenary, and so (11.1.1) holds.

(11.1.3) $\Rightarrow$ (11.1.4) $\Rightarrow$ (11.1.1) much as in the proof of (6.1.4) $\Rightarrow$ (6.1.5) $\Rightarrow$ (6.1.1) (but use (A.11.1) $\Rightarrow$ (A.11.7) and (3.1.7) in place of (1.3.5) and (3.1.4), respectively).

(11.1.3) $\Rightarrow$ (11.1.5), by (A.10.1) $\Rightarrow$ (A.10.10), (11.1.5) $\Rightarrow$ (11.1.11), by (A.4.1) $\Rightarrow$ (A.4.7), and (11.1.11) $\Rightarrow$ (11.1.12), by (A.3.1) $\Rightarrow$ (A.3.5).

(11.1.12) implies that if R is a catenary local domain, then, for all height one $p \in$ Spec R , R/p is a C_o-domain, by (A.4.7) $\Rightarrow$ (A.4.1), hence R is a C_1-domain, by (A.4.5) $\Rightarrow$ (A.4.1), and so (11.1.5) holds.

To show that (11.1.5) $\Rightarrow$ (11.1.3), it suffices, by [RP, (3.13)], to prove that (11.1.5) implies a catenary local domain is a C_i-domain, for $i = 1,\dots,$ altitude $R - 2$. For this, assume R is a C_i-domain, for some i ($1 \leq i <$ altitude $R - 2$) and let p be a height i prime ideal in R . Then R/p is catenary, by (1.2.1), so R/p is a C_1-domain, by (11.1.5), and so R is a C_{i+1}-domain, by (A.4.5) $\Rightarrow$

(A.4.1). Therefore, it follows that (11.1.5) $\Rightarrow$ (11.1.3).

Assume (11.1.3) holds and let A be a taut finite integral extension domain of a local domain R . Then R is catenary, by [McR-2,
Propositions 12 and 7], so A is catenary, by hypothesis and (A.10.1)
$\Rightarrow$ (A.10.5) (since a semi-local ring that satisfies the o.h.c.c. is
catenary, by (1.4.1)), hence (11.1.6) holds.

Assume (11.1.6) holds and let R be a catenary local domain, so
R is taut. Then every finite integral extension domain A of R is
taut, by (A.8.1) $\Rightarrow$ (A.8.3), so A is catenary, by (11.1.6), and so
(11.1.8) holds, by (1.2.1).

(11.1.3) $\Rightarrow$ (11.1.9), by (A.10.1) $\Rightarrow$ (A.10.12).

Assume (11.1.9) holds, let R be a catenary local domain, and
let p be a height one prime ideal in R . Let P be a minimal prime
divisor of pR^H , so height P = 1 , hence depth P = altitude R - 1 ,
by (11.1.9). Therefore, each maximal ideal in (R/p)' has height =
altitude R - 1 (by [N-6, Ex. 2, p. 188], since $R^H/pR^H \cong (R/p)^H$) , so
(11.1.5) holds.

(11.1.3) $\Leftrightarrow$ (11.1.10), as in the proof that (6.1.4) $\Leftrightarrow$ (6.1.10).

(11.1.11) $\Rightarrow$ (11.1.13), by (A.3.1) $\Rightarrow$ (A.3.5).

Finally, assume (11.1.13) holds and let p be a height one prime
ideal in a catenary local domain R . Then to prove that (11.1.5)
holds it may be assumed that (R/p)' is not quasi-local. Let N be
a maximal ideal in (R/p)' , and let d,e in M/p such that $e/d \in N$
and 1 - (e/d) is in all other maximal ideals in (R/p)' . Let b
and c be preimages in R of d and e , respectively, so (R/p)[e/d]
$\cong R[c/b]/p^*$ (where $p^* = pR[1/b] \cap R[c/b]$) and $p^* \subseteq (M,c/b)$ (since
(M/p,e/d) = $N \cap (R/p)[e/d]$ is proper). Now it may be assumed that
height (b,c)R = 2 , by the proof of [P, (4.7), pp. 49-50]. Therefore
height (M,c/b) = altitude R , by [R-9, Lemma 2.7], so height $(M,c/b)/p^*$
= altitude R - 1 , by (11.1.13). Thus, it follows that height N =

altitude $R - 1 = $ depth p , so R is a C_1-domain, hence (11.1.5) holds, q.e.d.

In [P, (8.8)], it was proved that (11.1.1) ⇔ (11.1.10) and that (11.1.1) is equivalent to (11.1.12) and to (11.1.13), both with "an H-domain" replaced by "catenary." Also, an additional (quite technical) equivalence to (11.1) is given in this same result in [P].

This chapter will be closed by showing that this conjecture could have been stated for an arbitrary catenary Noetherian domain.

(11.2) THEOREM. The following statements are equivalent:

(11.2.1) The Catenary Chain Conjecture (11.1.1) holds.

(11.2.2) If A is a catenary Noetherian domain, then A' is catenary.

(11.2.3) If A is a catenary Noetherian domain, then A' satisfies the c.c.

Proof. (11.2.3) ⇒ (11.2.2), by (1.3.5), and (11.2.2) ⇒ (11.2.1), by (11.1.2) ⇒ (11.1.1). Therefore assume that (11.2.1) holds, let A be a catenary Noetherian domain, let M' be a maximal ideal in A' , and let $M = M' \cap A$. Then A_M is catenary, by (1.2.1), so $A'_{(A-M)} = (A_M)'$ satisfies the c.c., by hypothesis. Now $N = M'A'_{(A-M)}$ is proper and $A'_{M'} = (A'_{(A-M)})_N$, so $A'_{M'}$ satisfies the c.c., by (1.3.1). Therefore A' satisfies the c.c., by (1.3.2), and so (11.2.3) holds, q.e.d.

(12.1.3) If R is a catenary local domain such that R' is quasi-local, then R satisfies the s.c.c.

(12.1.4) If R is a local domain such that R' satisfies the f.c.c., then every free quadratic integral extension domain of R satisfies the f.c.c.

Proof. Assume (12.1.1) holds and let R be as in (12.1.2). If R' is level, then R' satisfies the f.c.c., so R satisfies the s.c.c., by (12.1.1), hence R' does by (A.11.1) ⇒ (A.11.7). Therefore, assume R' is not level, so altitude R > 1 and there exists a height one maximal ideal in R', by hypothesis. Let b, S = R', B, and C be as in (3.1), so B = R[b,1/b] is a local domain and B' = C = R'[1/b] is taut-level and catenary, by (3.1.2), (3.1.5), and (3.1.6). Therefore it follows from what has already been proved that B' satisfires the s.c.c., so R' satisfies the c.c., by (3.1.7), and so (12.1.2) holds.

Assume (12.1.2) holds and let R be as in (12.1.3). Then R' is catenary, by (A.9.1) ⇒ (A.9.3), and R' is level, so R' is catenary and taut. Therefore R' satisfies the c.c. (by (12.1.2)) and the f.c.c., so R' satisfies the s.c.c., by (1.3.3). Hence R satisfies the s.c.c., by (1.3.4), and so (12.1.3) holds.

Assume (12.1.3) holds and let R be a local domain such that R' satisfies the f.c.c. Let A be a finite R-algebra such that R ⊆ A ⊆ R' and A and R' have the same number of maximal ideals. Then, for each maximal ideal N in A, $(A_N)' = R'_{(A-N)}$ is quasi-local and satisfies the f.c.c., by (1.2.1), so A_N is catenary, by (A.9.2) ⇒ (A.9.1), hence A_N satisfies the s.c.c., by (12.1.3). Therefore $R'_{(A-N)}$ satisfies the s.c.c., by (A.11.1) ⇒ (A.11.7), so it follows from (1.3.3) that R' satisfies the s.c.c. Hence R satisfies the s.c.c., by (1.3.4), and so (12.1.1) holds.

Finally, it is clear that (12.1.1) ⇒ (12.1.4), so assume that

CHAPTER 12

THE NORMAL CHAIN CONJECTURE

This conjecture arose from M. Nagata's incomplete proof of Proposition 1a in [N-3], in 1956. A related Proposition 1b also has an incomplete proof, and both results were repeated in 1959, in [N-5, pp. 85-86], and a corrected version of these results was given in 1962, in [N-6, (34.3)]. (Since the reference to [N-6, (33.10)] in the proof of [N-6, (34.3)] was not explained, this result was reproved and sharpened in [R-2, Theorem 3.11].)

One equivalence of this conjecture follows easily from [R-2, Corollary 3.12] (see (12.1.1) ⇔ (12.1.4)), and it was noted in the comment following (2.3.4) in [R-10] that this conjecture holds if the following condition holds: if there exists a mcpil n in an integral extension domain B of a Noetherian domain A , then there exists a mcpil n in B∩A' .

It follows from comparing (12.1.1) ⇔ (12.1.3) with (11.1.1) ⇔ (11.1.4) that this conjecture is very closely related to the Catenary Chain Conjecture. Also, it is shown in (14.1) that this conjecture holds for Henselian local domains and local domains of the form $L[X]_{(M,X)}$.

(12.1) gives three equivalences of this conjecture and the reader should refer to (A.8), (A.9), and (A.11) to obtain some additional equivalences.

(12.1) THEOREM. The following statements are equivalent:

(12.1.1) The Normal Chain Conjecture (3.3.9) holds: if the integral closure of a local domain R satisfies the f.c.c., then R satisfies the s.c.c.

(12.1.2) If R is a local domain such that R' is taut and catenary, then R' satisfies the c.c.

(12.1.4) holds, let R be a local domain such that R' satisfies the f.c.c., and let R[c] be a quadratic integral extension domain of R If c ∈ R' , then R[c] satisfies the f.c.c., by (1.2.4). If c ∉ R', then there are no linear polynomials in K = Ker (R[X] → R[c]) and X^2+rX+s ∈ K , for some r,s ∈ R (since c is quadratic over R) . Therefore, by the Division Algorithm, it follows that K = $(X^2+rX+s)R[X]$, so R[c] is a free R-algebra, hence R[c] satisfies the f.c.c., by hypothesis. Therefore R satisfies the s.c.c., by (A.11.9) ⇒ (A.11.1), and so (12.1.1) holds, q.e.d.

(The proof that (12.1.4) ⇒ (12.1.1) is essentially the same as that given in [R-2, Corollary 3.12].)

It follows from (12.1.4) that if "catenary" is inherited by flat finite integral extension domains, then this conjecture holds.

The final result in this chapter shows that this conjecture could have been stated for a more general case.

(12.2) PROPOSITION. The Normal Chain Conjecture (12.1.1) is equivalent to (*): if A is a Noetherian domain such that A' satisfies the f.c.c., then A satisfies the s.c.c.

Proof. It is clear that (*) ⇒ (12.1.1), so assume that (12.1.1) holds and let A be a Noetherian domain such that A' satisfies the f.c.c. Then A satisfies the f.c.c., by (1.2.4), so to prove that A satisfies the s.c.c. it suffices to prove that A_M satisfies the c.c., for all maximal ideals M in A , by (1.3.3). For this, $(A_M)'$ = $A'_{(A-M)}$ is level (since A' is) and catenary (by (1.2.1)), so $(A_M)'$ satisfies the f.c.c., by (1.2.3). Therefore A_M satisfies the s.c.c., by (12.1.1), and so (*) holds, q.e.d.

CHAPTER 13

COMMENTS ON (3.3.1) AND CONJECTURE (K)

I think that all the conjectures mentioned in Chapter 3, with the
exception of (3.3.1), hold, but I have strong doubts about (3.3.1).
The reasons for these doubts have to do with a question asked by I.
Kaplansky and mentioned by M. Hochster in [Hoc, p. 67]. Stated in the
terminology of [McR-1], the question becomes:

(13.1) Conjecture (K). If P and Q are height two prime ideals
in a Noetherian domain A , then there exists a height one $p \in$ Spec A
such that $p \subset P \cap Q$.

It can be proved that if Conjecture (K) holds, then the Chain Conjecture implies (3.3.1), so (3.3.1) $\Leftrightarrow$ (3.3.2). However, it is known
(see [H], [Mc-3], or [Mc-5, Theorem 7]) that Conjecture (K) does not
hold. In particular, R. Heitmann's construction in [H] shows that given
an integer $n > 1$, there exists a regular semi-local domain $(R;M,N)$
such that height $M = n =$ height N and (0) is the only prime ideal
in R contained in $M \cap N$. It seems to me that it should be possible
to adjust this construction of Heitmann to obtain a noncatenary local
domain R such that R' is a regular domain with two maximal ideals
(say M' and N') such that $1 <$ height $M' <$ height N' (and such that
there are no nonzero prime ideals contained in $M' \cap N'$) . If this is
the case, then such an R would be a completely new example of the
failure of the chain problem of prime ideals (since M. Nagata's examples
for noncatenary R have infinitely many prime ideals contained in the
Jacobson radical of R'), and would show that (3.3.1) is false. At
present, I do not know if such an R can be constructed.

Finally, it is shown in (14.2) that this conjecture holds for local
domains of the form $L[X]_{(M,X)}$.

(13.2) gives one equivalence of (3.3.1).

(13.2) THEOREM. The following statements are equivalent:

(13.2.1) (3.3.1) holds: if R is a local domain such that, for all nonzero P ∈ Spec R , R/P satisfies the s.c.c., then R is catenary.

(13.2.2) If R is a local domain such that, for all nonzero P ∈ Spec R , R/P satisfies the s.c.c., then R is an H-domain.

Proof. It is clear that (13.2.1) ⇒ (13.2.2), so assume (13.2.2) holds and let R be a local domain such that R/P satisfies the s.c.c., for all nonzero P ∈ Spec R . Let P be a nonzero prime ideal in R , so there exists a height one p ⊆ Spec R such that p ⊆ P . Then height P/p + depth P/p = altitude R/p = altitude R - 1 , by hypothesis. Therefore height P + depth P = altitude R , hence R is catenary, by (A.9.5) ⇒ (A.9.1), and so (13.2.1) holds, q.e.d.

(13.3) REMARK. It follows quite readily from [N-6, Example 2, pp. 203-205] in the case m = 0 that the conclusion in (3.3.1) cannot be replaced by: R satisfies the s.c.c. (Concerning this, see (B.3.3).)

CHAPTER 14

SOME EXAMPLES

A number of examples are given in this chapter to show that some
of the named conjectures in Chapter 3 hold for certain classes of local
domains, and, on the other hand, that there are quasi-local domains for
which certain of these conjectures do not hold. We begin by showing
that a number of the conjectures hold for a Henselian local domain and
for a local domain of the form $R[X]_{(M,X)}$.

(14.1) EXAMPLE. The Upper Conjecture, the Catenary Chain Conjecture, and the Normal Chain Conjecture hold for Henselian local domains
and local domains of the form $D = R[X]_{(M,X)}$, where (R,M) is a local
domain.

Proof. By (1.5.3) ⇔ (1.5.4), there exists a mcpil n+2 in
$D[Y]_{(M,X,Y)}$ if, and only if, there exists a mcpil n+1 in D , so the
Upper Conjecture holds for D . Also, by (1.5.1) ⇔ (1.5.4), there
exists a mcpil n+1 in D if, and only if, there exists a mcpil n in
some integral extension domain of R . Therefore, since every integral
extension domain of a Henselian local domain is quasi-local, the Upper
Conjecture holds for Henselian local domains, by (A.5.2) ⇒ (A.5.1).

[R-3, Theorem 2.21] shows that D and Henselian local domains
are catenary if, and only if, they satisfy the s.c.c., so these rings
satisfy the Catenary Chain Conjecture, by (A.11.1) ⇒ (A.11.7).

Finally, for the Normal Chain Conjecture, this was shown in
(3.5.2), q.e.d.

(14.2) REMARK. (3.3.1) also holds for D as in (14.1).

Proof. If D/XD = R satisfies the s.c.c., then D satisfies the
f.c.c., by (A.11.1) ⇒ (A.11.19), so (3.3.1) holds for D , q.e.d.

If (3.3.1) holds for Henselian local domains, then the Chain Conjecture holds, by the proof of (3.3.1) $\Rightarrow$ (3.3.2).

(14.3) EXAMPLE. Henselian semi-local rings and semi-local domains of the form $D = R[X]_S$, where R is semi-local and $S = R[X] - \cup \{(M,X)R[X]$; M is a maximal ideal in R$\}$, satisfy the Taut-Level Conjecture.

Proof. If R is a taut-level semi-local Henselian ring, then, for each minimal prime ideal z in R , R/z is taut, by [McR-2, Proposition 5], and R/z is a Henselian local domain. Thus R/z satisfies the f.c.c., by (A.9.5) $\Rightarrow$ (A.9.1), and altitude R/z = altitude R , hence R satisfies the f.c.c.

Also, if D is taut-level, then, for each maximal ideal M in R , $R_M[X]_{MR_M[X]} = D_{MD}$ satisfies the s.c.c., by [McR-2, Proposition 9]. Therefore R_M satisfies the s.c.c., by [RMc, (2.15)], so $D_{(M,X)} = R_M[X]_{(MR_M,X)}$ satisfies the f.c.c., by [R-3, Theorem 2.21]. Also, height $(M,X)D$ = altitude D , so D satisfies the f.c.c., by (1.2.3), q.e.d.

An alternate proof that the Taut-Level Conjecture holds for D is given in Remark (b) preceding Corollary 5 in [Mc-2].

(14.4) REMARK. The following statements are readily verified:

(14.4.1) If R is a quasi-local domain such that either altitude R $\leq$ 1 or altitude R = 2 and R' is level, then (3.3.1) and all the named conjectures in Chapter 3, except the Upper Conjecture, hold for R . (Some hold vacuously.) And, if the Upper Conjecture is restated using (9.2.3), then the Upper Conjecture also holds for R . (For the (3.8.5) version of the Upper Conjecture, see (14.5).)

(14.4.2) If R is a complete local domain, then (3.3.1) and all the named conjectures in Chapter 3 hold for R .

We now show that there are quasi-local domains for which some of
the named conjectures in Chapter 3 do not hold. It has already been
noted that this is the case for: the Chain Conjecture, by [S] or [K];
the H-Conjecture, by [Fu-1] or W. Heinzer's example (see (14.6)); the
Descended GB-Conjecture and the GB-Conjecture, by [K]; and, the Taut-
Level Conjecture, by W. Heinzer's example.

(14.5) EXAMPLE. The Upper Conjecture does not hold for certain
quasi-local domains.

Proof. Let R be the quasi-local domain mentioned in the last
paragraph of Section 2 in [Sei], so altitude R = 1 and altitude R[X]
= 3 , hence the Upper Conjecture does not hold for R , q.e.d.

In (14.6) and (14.7), we use the fact [L, (3.1)] that each finite
partially ordered set with a minimum element and a maximum element is,
in fact, isomorphic to Spec R , for some quasi-local Bezout domain R .
Both examples are readily verified.

(14.6) EXAMPLE. (W. Heinzer [R-14, (4.29.1)].) Let S = {a ,
b , c , d , e , f , g , h} with $a \subset b \subset d \subset f \subset h$, $a \subset c \subset e \subset g \subset h$,
and $a \subset c \subset f \subset h$ (a regular hexagon split down a diagonal and then
one part of the cut slid on the other part to the former center). Then,
with R a quasi-local Bezout domain such that Spec R $\cong$ S , R does
not satisfy either the H-Conjecture or the Taut-Level Conjecture.

(14.7) EXAMPLE. Let S = {a , b , c , d , e , f} with $a \subset b \subset$
$c \subset e \subset f$ and $a \subset b \subset d \subset f$. Then, with R a quasi-local Bezout
domain such that Spec R $\cong$ S , the Strong Avoidance Conjecture, the
Avoidance Conjecture, the Depth Conjecture, the Weak Depth Conjecture,
and the H-Conjecture fail to hold for R .

I do not know any examples for which the Catenary Chain Conjecture
or the Normal Chain Conjecture fail to hold.

CHAPTER 15
SOME RELATED QUESTIONS

In this chapter, a number of questions that are related to the chain conjectures are briefly discussed. The first is actually another chain conjecture. It comes from a weakening of the Normal Chain Conjecture, and it could have been given in Chapter 3 as (3.3.10). However, it was decided to delay giving it until this point, since the main difficulty in settling this problem is essentially the same difficulty in showing that the Normal Chain Conjecture holds. In fact, it is this same difficulty that arises in (15.1a) - and if it can be shown that (15.1a) has an affirmative answer, then the same method of proof may well provide the needed information to show that (15.1) and (3.3.9) hold.

(15.1) If R is an integrally closed catenary local domain, then does R satisfy the s.c.c.?

(15.1a) is a very special case of (15.1). (R in (15.1a) is catenary, since altitude R = 3 and R is an H-domain (since height one prime ideals are principal).)

(15.1a) [R-12, (4.4)]. If R is a local UFD (unique factorization domain) such that altitude R = 3 , does R satisfy the s.c.c.?

It was noted in [R-19, (6.1)] that (15.1a) is equivalent to: is such a ring R a GB-domain?

R is catenary in (15.1a), so, if R is also Henselian, then, as noted by H. Seydi in [Sey, Corollary II.2.4], the answer is yes. (See [R-3, Theorem 2.21] for a more general result.) Also, [R-2, Corollary 3.12] shows that the answer is yes if all free quadratic integral extension domains of R are catenary.

Finally, in [Fu-2, Proposition], K. Fujita constructed a quasi-

local UFD R of altitude three that is not catenary, and so this R
does not satisfy the s.c.c.

(15.2) Three questions on semi-local UFD's.

(15.2.1) Are all local Henselian UFD's of altitude four catenary?

(15.2.2) Are all local UFD's of altitude four catenary?

(15.2.3) Are all semi-local taut-level UFD's of altitude four
catenary?

For all three questions, note that if R is the ring, then there
are no mcpil 2 in R (since height one prime ideals are principal)
and if P is a non-maximal prime ideal in R , the R_P is catenary
(since R_P is a UFD and altitude $R_P \leqq 3$) , so it is sufficient to
show that if height $M/P = 1$, then height $P = 3$.

(Note that (15.2.1) and (15.2.2) are special cases of the H-Conjecture, and (15.2.3) is a special case of the Taut-Level Conjecture.)

(15.3) [R-16, (6.1.3)]. If (R,M) is a local GB-domain, then
is $R(X) = R[X]_{MR[X]}$ a GB-domain?

It is known [R-19, (2.8) and (4.1)] that (15.3) is equivalent to
each of the following:

(a) Is such a ring R catenary?

(b) For such a ring R , is R[X] a GB-domain?

(c) For such a ring R , is $R[X]_N$ a GB-domain, for some maximal ideal N in R[X] such that $N \cap R = M$?

(15.4) Questions on saturated chains of prime ideals in integral
extension domains of a local domain (R,M) .

(15.4.1) [R-10, (2.3)]. If $(0) \subset Q_1 \subset \cdots \subset Q_n$ is a mcpil n
in some integral extension domain B of R , then:

(a) Is height Q_i = height $Q_i \cap (B \cap R')$?

(b) Is there a finite integral extension domain A of R that has a mcpil n , say $(0) \subset P_1 \subset \cdots \subset P_n$, such that height P_i = height Q_i and depth P_i = depth Q_i , for $i = 1,\ldots,n$?

(c) If so, can A be chosen as a subring of B and such that $P_i = Q_i \cap A$?

(d) Can A be chosen as a subring of $B \cap R'$?

(e) Can A in (b) be chosen as a principal integral extension domain of R ?

It was noted in [R-10, p. 79] that the answer to (15.4.1)(a) is yes, if R = R' , and that the answers to (15.4.1)(b) and (c) are yes, if B lies between a Noetherian domain and its integral closure. Also, for each $i = 1,\ldots,n$, there exists a principal integral extension do-main $R[b_i] \subset R'$ that has a prime ideal P_i such that height P_i = height Q_i and depth P_i = depth Q_i , by [R-4, Lemma 2.9].

(15.4.2) Does there exist a principal integral extension domain $R[c]$ of R such that there exists a mcpil n in some integral exten-sion domain of R (if and) only if there exists a mcpil n in $R[c]$?

Concerning (15.4.2), it is known [HMc, Theorem 1.10] that if there exists a mcpil n in some integral extension domain of R , then there exists a principal integral extension domain $R[b]$ of R that has a mcpil n .

(15.4.3) If $(0) \subset P_1 \subset \cdots \subset P_n$ is a mcpil n in R , are there infinitely many mcpil n in R , say $(0) \subset P_1' \subset \cdots \subset P_n'$, such that height P_i' = height P_i and depth P_i' = depth P_i , for $i = 1$, $\ldots,n$?

For (15.4.3), it is known that there exist infinitely many mcpil n in R , by [McR-2, Lemma 1] and its proof. It is also known [R-22] that there may be only finitely many choices for a given P_i . For example,

if n = 3 and height P_2 = 3 (and little height P_2 = 2) , then there
may be only finitely many prime ideals Q in R such that height
Q = 3 and little height Q = 2 .

(15.4.4) If R ⊆ S are quasi-local domains such that R is Noetherian and S is integral over R , and if there exists a mcpil n
in R , say (0) ⊂ P_1 ⊂ ⋯ ⊂ P_n , such that height P_i = h_i and depth
P_i = d_i , then does there exist such a chain in S ? Does the converse
hold?

Concerning (15.4.4), it is known [R-10, (3.1)] that there exists
a mcpil n in R if and only if there exists a mcpil n in S .

(15.4.5) [R-12, (4.6)]. If R has the property that there exists
a mcpil n in some integral extension domain of R (if and) only if
there exists a mcpil n in R , does every integral domain S_N inherit
this property, where S is integral over R and N is a maximal ideal
in S ? Does every locality over R inherit this property?

It is known [R-12, (4.1.2) and (4.1.3)] that Henselian local domains and $R[X]_{(M,X)}$ (where (R,M) is an arbitrary local domain) have
the property mentioned in (15.4.5). It is also known [R-12, (4.8.2)]
that if R has this property and R' is quasi-local, then each S_N
as above inherits this property from R .

(Note that the property in (15.4.5) is closely related to the
Upper Conjecture (3.8.5), by (1.5.1) ⟺ (1.5.4).)

(15.5) Does there exist a Noetherian Hilbert D_1-domain A that
is catenary and has a maximal ideal N such that 1 < height N < altitude A ?

If the answer to (15.5) is no, then the Catenary Chain Conjecture
holds. (For, if (R,M) is a catenary local domain, then $D = R[X]_{(M,X)}$

is a local D_2-domain, by (A.9.1) $\Rightarrow$ (A.9.16). Therefore, for each non-zero $f \in (M,X)D$, $D[1/f]$ is a D_1-domain, by [R-4, Lemma 2.1], and $D[1/f]$ is a Noetherian Hilbert domain. Also, by (1.2.2) and [R-4, Corollary 3.11 and Theorem 4.11], it can be proved that $D[1/f]$ is catenary. Now, if, for each maximal ideal N in $D[1/f]$, height $N \in \{1$, altitude $D[1/f] = $ altitude $R\}$, then it follows from [R-4, Lemma 2.1] (and since f is arbitrary) that D is an H_2-domain, so the Catenary Chain Conjecture holds, by (11.1.11) $\Rightarrow$ (11.1.1).)

(15.6) [R-3, Remark 5.12]. Questions on unmixedness of a local domain.

(15.6.1) If R satisfies the f.c.c. and $R^{(1)}$ is a finite R-algebra, is R quasi-unmixed?

(15.6.2) If R is quasi-unmixed and $R^{(1)}$ is a finite R-algebra, is R unmixed?

(15.6.3) If R is integrally closed, is R unmixed?

(15.6.4) See the questions mentioned in (2.3) and (2.4).

If $R^{(1)}$ is a finite R-algebra and altitude $R > 1$, then there does not exist a height one maximal ideal in R', by [R-3, Corollary 5.7(i)]. Therefore, if the Catenary Chain Conjecture holds, then the answer to (15.6.1) is yes, by (3.2), (11.1.1) $\Leftrightarrow$ (11.1.4), and (A.11.1) $\Leftrightarrow$ (A.11.4). Also, note that if $R = R'$, then $R^{(1)} = R$, so in this case it is not necessary to assume that $R^{(1)}$ is a finite R-algebra. Finally, it is known [R-3, Lemma 5.11(1)] that if R is unmixed, then $R^{(1)}$ is a finite R-algebra, and it is shown in (B.5.2) that the converse does not hold.

(15.7) (D. Rush) Does every Henselian local domain R have a separated maximal Cohen-Macaulay module M such that Supp $M = $ Spec R ?

Concerning (15.7), in [Hoc, p. 69], M. Hochster asked if maximal

Cohen-Macaulay modules with good properties can be constructed, and some affirmative results have recently been obtained. One is given in [Ru, Corollary 2]: if R is a complete equicharacteristic local ring that has a maximal Cohen-Macaulay module as in (15.7), then R satisfies the f.c.c. If this continues to hold for Henselian local domains, then by (4.1.1) ⇔ (4.1.3), the Chain Conjecture is translated into a problem in homological algebra by (15.7). Since many results are currently being discovered in the area of Big Cohen-Macaulay modules, and since powerful homological methods are readily available, an affirmative answer to this question may go a long way toward settling the Chain Conjecture.

APPENDIX A

A SUMMARY OF KNOWN EQUIVALENCES FOR CERTAIN CHAIN CONDITIONS

In this Appendix we list a number of characterizations of nine of
the chain conditions considered in these notes. These lists can be
used to give formally different characterizations of the various characterizations of the chain conjectures given in Chapters 4 - 13, and they
can be used in like manner for the other chain conjectures mentioned
in Chapter 3.

Some of the lists are quite long, but in no case is the list ex-
haustive - that is, quite a few other such characterizations of these
conditions have appeared in the literature. However, the lists do in-
clude at least most of the more important characterizations, and they
indicate the variety of the characterizations of each condition that
has appeared in the literature. (A few of the characterizations are,
in fact, new, but these follow quite readily from previously known characterizations.) In the lists an attempt has been made to group the
statements that are concerned with similar concepts near to each other.
Also, an effort has been made to keep approximately the same order of
appearance of similar concepts in the different lists. Some of the
characterizations are weak, in the sense that they should be easier to
verify than the definition. Others are strong, in the sense that they
show some unexpected things hold, if the characterized condition holds.

Before giving the lists, we need the following notation and lemma.

(A.1) NOTATION. Let (R,M) be a local ring and let $b,c_1,\ldots,c_k$
be analytically independent elements in R , so b is not nilpotent.
Let $Z = (0):b^iR$, for all large i , let $^\circ$ denote residue class
modulo Z , and let $A = (R/Z)[c_1^\circ/b^\circ,\ldots,c_k^\circ/b^\circ]$. Then $(M/Z)A$ is
a depth k prime ideal, by [R-4, Remark 4.4(i)], so we denote the lo-
cal ring $A_{(M/Z)A}$ by $R(c_1/b,\ldots,c_k/b)$.

The following lemma will help clarify the applications of (A.3.1) ⇔ (A.3.7) and (A.3.1) ⇔ (A.3.9) in Chapter 4.

(A.2) LEMMA. If (R,M) is a local domain and b,c are in M , then b,c are analytically independent in R if and only if there exists a maximal ideal M' in R' such that c/b and b/c are not in $R'_{M'}$.

Proof. By [R-3, Lemma 4.3], b,c are analytically independent in R if and only if MR[c/b] is a depth one prime ideal. Now this holds if and only if M'R'[c/b] is a depth one prime ideal, for some maximal ideal M' in R' , by integral dependence, and this is equivalent to c/b and b/c are not in $R'_{M'}$, by [ZS-2, Corollary, p. 20], q.e.d.

Before each of the next nine remarks a summary of the results in Chapters 3 - 13 in which the condition has appeared is given. (In some cases a secondary list is included in brackets [ ] listing results for which the characterized condition is only partly applicable - such as R' satisfies the c.c., for (A.11).

For (A.3), see: (3.3.6), (3.3.7), (3.9.4), (3.10.2), (3.10.3), (3.11.2), (3.11.3), (3.12), (4.1.4), (4.1.5), (4.1.9), all of (6.1), (6.3), (11.1.8), (11.1.9), (11.1.10), (11.1.11), (11.1.12), (11.1.13), and (13.2.2). [See also: (3.1.4), (3.2.1), (3.9.2), (3.9.3), (3.10.1), (3.14.2), and (11.1.7).]

(A.3) REMARK. The following statements are equivalent for a local ring (R,M) , where a = altitude R :

(A.3.1) R is an H_i-ring.

(A.3.2) There exists S in $\mathcal{S}$ such that S is an H_i-ring, where $\mathcal{S}$ = {S ; S is a quasi-local integral extension ring of R such that, for each minimal prime ideal z in S , either z∩R is minimal or depth z ≤ i} .

(A.3.3) Every S in $\mathfrak{s}$ is an H_i-ring, where $\mathfrak{s}$ is as in (A.3.2).

(A.3.4) $R/(\text{Rad } R)$ is an H_i-ring.

(A.3.5) For each fixed j ($0 \leq j \leq i$) and for each $p \in \text{Spec } R$ such that height $p = j$, R/p is an H_{i-j}-ring and either depth $p = a-j$ or depth $p \leq i-j$.

(A.3.6) Every set of $i+1$ analytically independent elements in R can be extended to a set of a analytically independent elements in R.

(A.3.7) For each fixed k ($0 \leq k \leq i$) and for all analytically independent elements $b, c_1, \ldots, c_k$ in R, $R(c_1/b, \ldots, c_k/b)$ (see (A.1)) is an H_{i-k}-ring and its altitude is either $= a-k$ or is $\leq i-k$.

(A.3.8) $R(X) = R[X]_{MR[X]}$ is an H_i-ring.

If $i = 1$ and R is an integral domain, then these statements are also equivalent to the following statements:

(A.3.9) For all analytically independent elements b, c in R, altitude $R[c/b] = a$.

(A.3.10) For all elements u in an algebraic closure F^* of the quotient field of R such that $MR[u] \neq R[u]$ and for all maximal ideals N in $R[u]$ such that $MR[u] \subseteq N$, height $N \in \{1, a\}$.

(A.3.11) For all $u \in F^*$ as in (A.3.10), if $(M, u)R[u] \neq R[u]$, then height $(M, u)R[u] \in \{1, a\}$.

If, moreover, either $a = 1$ or there are no height one maximal ideals in R', then these statements are also equivalent to the following statements:

(A.3.12) R' is an H-domain.

(A.3.13) There exists a finite algebraic extension domain S of R such that altitude $S = a$ and S' is an H-domain.

(A.3.14) With u, F^*, and N as in (A.3.10), height $N = a$.

(A.3.15) With u and F^* as in (A.3.11), height $(M, u)R[u] = a$.

(A.3.16)   Altitude $R[u] = a$, for all   u   in the quotient field
of   R   such that $1/u \notin J'$, the Jacobson radical of   R' .

(A.3.17) $D = R[X]_{(M,X)}$ is an   H-domain.

(A.3.18)   If $p \subseteq \operatorname{Spec} R[X]_{(M,X)}$ has height one and contains a
linear polynomial, then   depth $p = a$.

(A.3.19) $R' = \cap\{V ; (V,N) \in \mathcal{V}\}$, where $\mathcal{V}$ is as in (4.2.2).

(A.3.20) $J' = \cap\{N ; (V,N) \in \mathcal{V}\}$, where $\mathcal{V}$ is as in (4.2.2) and
J'   is the Jacobson radical of   R' .

Proof.   (A.3.1) - (A.3.8) are all equivalent, by [RP, Section 2
and (3.15)].

(A.3.1), (A.3.9) - (A.3.11), and (A.3.13) are all equivalent, by
[R-21, (9.12), (9.13), and (9.5)].

(A.3.1) $\Leftrightarrow$ (A.3.12) $\Leftrightarrow$ (A.3.17), by [R-6, (3.2)].

Finally, (A.3.1), (A.3.14) -(A.3.16), and (A.3.18) - (A.3.20) are
all equivalent, by [R-17, (2.11), (2.8), (2.9), (2.6), and (2.7)], q.e.d.

For (A.4), see: (6.2.1), (6.2.2), and (11.1.5).   [See also
(3.2.5).]

(A.4)   REMARK.   The following statements are equivalent for a local ring   (R,M) , where $a =$ altitude R :

(A.4.1)   R   is a C_i-ring.

(A.4.2)   There exists $S \in \mathcal{S}$ such that   S   is a C_i-ring, where
$\mathcal{S}$ is as in (A.3.2).

(A.4.3)   Every $S \in \mathcal{S}$ is a C_i-ring, where $\mathcal{S}$ is as in (A.3.2).

(A.4.4) $R/(\operatorname{Rad} R)$ is a C_i-ring.

(A.4.5)   For each fixed   j   ($0 \leqq j \leqq i$)   and for each $p \in \operatorname{Spec} R$
such that   height $p = j$, R/p is a C_{i-j}-ring and either   depth $p =$
a-j   or   depth $p \leqq i-j$.

(A.4.6) $R(X) = R[X]_{MR[X]}$ is a C_i-ring.

(A.4.7) $R[X]_{(M,X)}$ is an H_{i+1}-ring.

(A.4.8)   For all $b \in M$ such that   height $bR = 1$, $\mathcal{L}(R,bR)$ is an H_{i+1}-ring, where $\mathcal{L}(R,bR) = \aleph_m$ with $\aleph = \aleph(R,bR)$ the Rees ring of   R   with respect to   bR   and $m = (M,u,tb)\aleph$.

(A.4.9)   For all   b   as in (A.4.8), $\mathfrak{Z}_n$ is an H_i-ring, where $\mathfrak{Z}$ = $\mathfrak{Z}(R,bR)$ is the associated graded ring of   R   with respect to   bR   and   n   is the maximal homogeneous ideal in $\mathfrak{Z}$.

If   R   is an integral domain, then these statements are also equivalent to:

(A.4.10)   There exists   k $(1 \le k < i)$ such that, for all analytically independent elements $b,c_1,\ldots,c_k$ in   R , $R(c_1/b,\ldots,c_k/b)$ (see (A.1))   is a C_{i-k}-ring.

(A.4.11)   For all   k $(1 \le k \le i)$ and for all analytically independent elements $b,c_1,\ldots,c_k$ in   R , $R(c_1/b,\ldots,c_k/b)$ (see (A.1)) is a C_{i-k}-ring.

(A.4.12)   For each   y   in the quotient field of   R   such that $(M,y)R[y] \neq R[y]$, $R[y]_{(M,y)}$ is an H_i-ring and its altitude is either = a   or is $\le i$.

Proof.   (A.4.1) - (A.4.5), (A.4.7), and (A.4.12) are all equivalent, by [RP, Section 3].

(A.4.1) $\Leftrightarrow$ (A.4.6) $\Leftrightarrow$ (A.4.10) $\Leftrightarrow$ (A.4.11), by [R-15, (4.1) and (4.2)]; and, (A.4.1) $\Leftrightarrow$ (A.4.8) $\Leftrightarrow$ (A.4.9), by [R-9, (3.1.2), (2.14.3), and (2.15)], q.e.d.

For (A.5), see: (3.4.1), (3.4.2), (3.8.5), (3.14.3), (4.1.7), (4.1.10), (5.3.2), (5.4), (7.2), all of (9.1), of (9.2), and of (9.3). [See also: (3.8.2), (3.8.3), (3.8.4), (3.14.1), (3.14.4), (5.1.2), (5.2), (8.1.2), and (8.3.2).]

(A.5)   REMARK.   The following statements are equivalent for a local domain   (R,M) :

(A.5.1)   There exists a   mcpil n   in   R .

(A.5.2) There exists a quasi-local integral extension domain S of R such that there exists a mcpil n in S .

(A.5.3) Every integral extension domain S of R has a mcpil n .

(A.5.4) There exists P $\in$ Spec R such that height P = n-1 and depth P = 1 .

(A.5.5) There exists an integral extension domain of R that has a depth one prime ideal P such that height P = n-1 .

(A.5.6) Every integral extension domain of R has a prime ideal P as in (A.5.5).

(A.5.7) There exists a mcpil n in $R(X) = R[X]_{MR[X]}$.

(A.5.8) There exists a quasi-local integral extension domain (S,N) of R such that there exists a mcpil n in $S(X) = S[X]_{NS[X]}$.

(A.5.9) Every quasi-local integral extension domain (S,N) of R is such that there exists a mcpil n in $S(X) = S[X]_{NS[X]}$.

Proof. (A.5.1), (A.5.2), and (A.5.7) - (A.5.9) are all equivalent, by [R-10 (3.1) and (3.10)].

(A.5.1) $\Rightarrow$ (A.5.3), by the Going Up Theorem, and it is clear that (A.5.3) $\Rightarrow$ (A.5.1).

(A.5.1) $\Rightarrow$ (A.5.4), by [Mc-1, Theorem 5], and clearly (A.5.4) $\Rightarrow$ (A.5.1).

Finally, it is clear that (A.5.6) $\Rightarrow$ (A.5.4) $\Rightarrow$ (A.5.5), and (A.5.5) $\Rightarrow$ (A.5.6), by [R-10, (2.8)], q.e.d.

For (A.6), see: (3.6.2), (3.6.3), (3.6.4), (3.7), all of (7.1), (7.2), (7.3), all of (7.4), and (7.5).

(A.6) REMARK. The following statements are equivalent for a ring A :

(A.6.1) A is a GB-ring.

(A.6.2) Every integral extension ring of A is a GB-ring.

(A.6.3) A/z is a GB-ring, for each minimal prime ideal z in A .

(A.6.4) A/I is a GB-ring, for each ideal I in A .

(A.6.5) A_M is a GB-ring, for each maximal ideal M in A .

(A.6.6) A_S is a GB-ring, for each multiplicatively closed subset S $(0 \notin S)$ in A .

If $A = R$ is a local domain, then these statements are also equivalent to the following statements:

(A.6.7) Every principal integral extension domain $R[c]$ of R is such that all pairs of adjacent prime ideals in $R[c]$ contract in R to adjacent prime ideals.

(A.6.8) R/p is a GB-ring, for each height one $p \in \operatorname{Spec} R$, and $R^{(1)} \subseteq R'$.

(A.6.9) For all $p \subset q$ in $\operatorname{Spec} R$, if there exists a mcpil n in some integral extension domain of $L = (R/p)_{q/p}$, then there exists a mcpil n in L .

(A.6.10) For all $p \subset q$ in $\operatorname{Spec} R$ such that height $q/p > 1$, there does not exist a height one maximal ideal in $((R/p)_{q/p})'$.

Proof. (A.6.1) - (A.6.6) are all equivalent, by [R-16, (3.1), (3.3), and (3.5)].

(A.6.1) $\Leftrightarrow$ (A.6.7) $\Leftrightarrow$ (A.6.9), by [R-16, (4.2) and (4.3)]; and (A.6.1) $\Leftrightarrow$ (A.6.8) $\Leftrightarrow$ (A.6.10), by [R-19, (5.3) and (5.2)], q.e.d.

For (A.7), see: (6.1.9) and (10.1.6).

(A.7) REMARK. The following statements are equivalent for a local ring (R,M) , where a = altitude R :

(A.7.1) R is taut.

(A.7.2) There exists an integral extension ring S of R such that minimal prime ideals in S lie over minimal prime ideals in R and such that S is taut.

(A.7.3) Every integral extension ring S of R such that minimal prime ideals in S lie over minimal prime ideals in R is taut.

(A.7.4) R/(Rad R) is taut.

(A.7.5) For all P $\in$ Spec R such that depth P $\leqq$ 1 , height P + depth P $\in$ {1,a} .

(A.7.6) R is an H_i-ring, for i = 1,...,a-2 .

(A.7.7) Every maximal set of analytically independent elements in R has either one element or a elements.

(A.7.8) For all analytically independent elements b,c in R , R(c/b) (see A.1)) satisfies the f.c.c. and altitude R(c/b) = a-1 .

(A.7.9) For all analytically independent elements $b,c_1,...,c_k$ in R (1 $\leqq$ k < a) , $R(c_1/b,...,c_k/b)$ (see (A.1)) satisfies the f.c.c. and has altitude = a-k .

(A.7.10) R(X) = $R[X]_{MR[X]}$ is taut.

(A.7.11) For all b $\in$ M such that height bR = 1 , every homogeneous prime ideal p in $\mathcal{R}$ such that height p > 1 is such that height $\mathfrak{m}$/p = a + 1 - height p , where $\mathcal{R}$ = $\mathcal{R}$(R,bR) and $\mathfrak{m}$ are as in (A.4.8).

Proof. (A.7.1) - (A.7.3) are equivalent, by [McR-2, Proposition 12].

(A.7.1), (A.7.4), and (A.7.6) - (A.7.10) are all equivalent, by [RP, (3.15), (2.3.4), (2.16.2), (2.14.2), (2.14.3), and (2.8.2)].

(A.7.1) $\Leftrightarrow$ (A.7.5), by [R-14, (2.13.1)]; and, (A.7.1) $\Leftrightarrow$ (A.7.11), by [R-9, (3.18.1)], q.e.d.

For (A.8), see: (3.1.5), (3.9.4), (3.9.5), (3.10.3), (3.14.2), (3.14.4), (6.1.3), all of (10.1), and (11.1.6). [See also: (6.1.2) and (12.1.2).]

(A.8) REMARK. The following statements are equivalent for a semilocal domain R , where a = altitude R :

(A.8.1) R is taut.

(A.8.2) There exists a taut integral extension domain of R .

(A.8.3) Every integral extension domain of R is taut.

(A.8.4) For each depth one P ∈ Spec R , height P = a-1 .

(A.8.5) For each non-maximal P ∈ Spec R , height P + depth P = a and R_P satisfies the s.c.c.

(A.8.6) For each non-zero b in the Jacobson radical of R , R_b satisfies the s.c.c.

(A.8.7) There exists a non-zero b in the Jacobson radical of R such that R_b satisfies the f.c.c.

(A.8.8) For all analytically independent elements b,c in R (that is, b,c are analytically independent in R_M , for some maximal ideal M in R) , A_S is taut and altitude A_S = a-1 , where A = R[c/b] and S is the complement in A of the ideals MA with M a maximal ideal in R such that b,c are analytically independent in R_M .

(A.8.9) For all k = 1,...,z-2 and for all analytically indepen-dent elements $b,c_1,...,c_k$ in R (see (A.8.8)), altitude A_S = a-k , where A = $R[c_1/b,...,c_k/b]$ and S is the complement in A of the ideals MA with M a maximal ideal in R such that $b,c_1,...,c_k$ are analytically independent in R_M .

(A.8.10) R(X) = $R[X]_S$ is taut, where S = R[X] - U{MR[X] ; M is a maximal ideal in R} .

Proof. (A.8.1) - (A.8.3) are equivalent by [McR-2, Proposition 12]; and, (A.8.1) and (A.8.4) - (A.8.10) are all equivalent, by [R-14, (2.13.2), (3.10), (3.21), (4.17), (4.20), and (4.10.1)], q.e.d.

For (A.9), see: (3.3.1), (3.3.6), (3.3.8), (3.5.1), (3.7), (3.12), (4.1.3), (6.1.8), (10.1.6), all of (11.1), (12.1.3), (12.1.4), and (13.2.1). [See also: (3.1.6), (3.3.9), (3.5.2), (3.9.2), (3.9.5), (3.10.1), (4.4.2), (10.1.1), (10.1.2), all of (11.2), (12.1.2), and (12.2).]

(A.9)   REMARK.   The following statements are equivalent for a local domain   (R,M) , where   a = altitude R :

(A.9.1)   R is catenary (equivalently, satisfies the f.c.c.).

(A.9.2)   There exists a quasi-local integral extension domain of R that is catenary.

(A.9.3)   All quasi-local integral extension domains of R are catenary.

(A.9.4)   For all $P \in$ Spec R   such that   depth P = 1 , height P + depth P = a .

(A.9.5)   For all $P \in$ Spec R , height P + depth P = a .

(A.9.6)   With at most finitely many exceptions, if $p \in$ Spec R and   height p = 1 , then   depth p = a - 1 and R/p is catenary.

(A.9.7) R_b satisfies the s.c.c., for all $0 \neq b \in M$.

(A.9.8) R_b satisfies the f.c.c., for some $0 \neq b \in M$.

(A.9.9) R_b is level, for some $0 \neq b \in M$.

(A.9.10)   For all systems of parameters $x_1,\ldots,x_a$ in R , height $(x_1,\ldots,x_i)R = i$, for i = 1,...,a .

(A.9.11)   Each maximal set of analytically independent elements in R contains a elements.

(A.9.12)   For all analytically independent elements b,c in R , R(c/b) (see (A.1)) is catenary and altitude R(c/b) = a - 1 .

(A.9.13)   For each k = 1,...,a-1 and for each set of analytically independent elements $b,c_1,\ldots,c_k$ in R , $R(c_1/b,\ldots,c_k/b)$ (see (A.1)) is catenary and altitude $R(c_1/b,\ldots,c_k/b) = a-k$.

(A.9.14)   For each k = 1,...,a-2 , altitude $R(c_1/b,\ldots,c_k/b) =$ a-k (see (A.1)).

(A.9.15) $R(X) = R[X]_{MR[X]}$ is catenary.

(A.9.16)   Every depth two prime ideal in $D = R[X]_{(M,X)}$ has height = a - 1 .

Proof.   (A.9.1) - (A.9.3), (A.9.7), (A.9.8), (A.9.11), (A.9.12), and (A.9.15), are all equivalent by the main theorem in [R-4].

(A.9.1) ⇔ (A.9.4) ⇔ (A.9.5), by [R-5, Theorem 2.2 and Remark 2.6(i)]; (A.9.1) ⇔ (A.9.6) ⇔ (A.9.10), by [R-4, Lemma 3.16] and [R-5, Remark 2.6(ii)], respectively; (A.9.1) ⇔ (A.9.9), by [R-11, (2.21)]; (A.9.1) ⇔ (A.9.13) ⇔ (A.9.14), by [R-4, (4.15)]; and, (A.9.1) ⇔ (A.9.16), by [Hou-1, Corollary 13], q.e.d.

For (A.10), see: (6.1.4) and (11.1.3). [See also (10.1.4).]

(A.10) REMARK. The following statements are equivalent for a semi-local domain R, where a = altitude R :

(A.10.1) R satisfies the o.h.c.c.

(A.10.2) R is taut and all finitely generated algebraic extension domains of R are catenary.

(A.10.3) R' satisfies the o.h.c.c.

(A.10.4) There exists an integral extension domain of R that satisfies the o.h.c.c.

(A.10.5) All integral extension domains of R satisfy the o.h.c.c.

(A.10.6) R is taut and all principal integral extension domains $R[c]$ of R are catenary.

(A.10.7) R is taut and all integral extension domains $R[c]$ of R such that $c^2 - rc \in R$, for some $r \in R$, are catenary.

(A.10.8) For each maximal ideal M in R, R_M satisfies the o.h.c.c. and altitude $R_M \in \{1, a\}$.

(A.10.9) R is taut and, for all nonzero $P \in$ Spec R such that depth $P > 1$, there does not exist a height one maximal ideal in $(R/P)'$.

(A.10.10) R is taut and R/P satisfies the s.c.c., for all nonzero $P \in$ Spec R.

(A.10.11) If there exists a mcpil n in an integral extension domain of R, then $n \in \{1, a\}$.

(A.10.12) R^H is taut.

(A.10.13) R^H satisfies the o.h.c.c.

(A.10.14) R is taut and depth P = depth P∩R , for all nonminimal P ∈ Spec R^H .

(A.10.15) R^* is taut.

(A.10.16) R^* satisfies the o.h.c.c.

(A.10.17) R(X) satisfies the o.h.c.c., where R(X) = R[X] - U{MR[X] ; M is a maximal ideal in R} .

(A.10.18) For each maximal ideal M in R , height M ∈ {1,a} and there exists i (1 ≦ i ≦ a-2) such that $R_M[X_1,\ldots,X_i]_{(M,X_1,\ldots,X_i)}$ is an H_j-domain, for j = i+1,...,a-1 .

(A.10.19) (A.10.18) holds with "there exists" replaced by "for all."

(A.10.20) For all y in the quotient field of R and for all maximal ideals M in R such that (M,y)R[y] ≠ R[y] , $A = R[y]_{(M,y)}$ is catenary and altitude A ∈ {1,a} .

(A.10.21) Every DVR (V,N) in the quotient field of R such that R ⊆ V, N∩R is a maximal ideal, and V is integral over a locality over R, is either of the first kind or $V = R'_{M'}$, for some height one maximal ideal M' in R' .

(A.10.22) For all ideals I in R such that depth I = 0 , all the minimal prime ideals in the associated graded ring of R with respect to I have depth ∈ {1,a} .

(A.10.23) For all ideals I in R such that depth I = 0 , all maximal relevant ideals in the Rees ring ℛ = ℛ(R,I) of R with respect to I have height ∈ {1,a} . (An ideal H in ℛ is a maximal relevant ideal in case H is homogeneous and is maximal with respect to not containing tI .)

Proof. These are all shown to be equivalent in [R-18, Section 4], q.e.d.

For (A.11), see: (3.3.1), (3.3.9), (3.5.2), (4.1.2), (6.1.5), (11.1.4), (12.1.3), all of (13.2), and (13.3). [See also; (3.1.7) (3.3.2), (3.3.8), (3.7), (4.4.3), (10.1.3), (10.1.5), (11.2.3), (12.1.2), and (12.2).]

(A.11)   REMARK.   The following statements are equivalent for a local domain (R,M) , where a = altitude R :

(A.11.1)   R   satisfies the s.c.c.

(A.11.2)   R   satisfies the altitude formula.

(A.11.3)   R   satisfies the dominating altitude formula.

(A.11.4)   R   is quasi-unmixed.

(A.11.5)   All finitely generated integral domains over   R   satisfy the c.c.

(A.11.6)   There exists an integral extension domain of   R   that satisfies the s.c.c.

(A.11.7)   All integral extension domains of   R   satisfy the s.c.c.

(A.11.8)   All principal integral extension domains $R[c]$ of   R satisfy the f.c.c.

(A.11.9)   All $R[c]$ as in (A.11.8)   such that $c^2 - rc \in R$, for some $r \in R$, satisfy the f.c.c.

(A.11.10)   R   is catenary and, for all $p \in$ Spec R   such that depth $p > 1$, there does not exist a height one maximal ideal in $(R/p)'$.

(A.11.11)   R   is a C_i-ring, for $i = 0,1,\ldots,a-2$.

(A.11.12)   R'   satisfies the f.c.c. and R/p satisfies the s.c.c., for all $(0) \neq p \in$ Spec R .

(A.11.13)   R   is catenary and is a   GB-domain.

(A.11.14) R^H satisfies the f.c.c.

(A.11.15)   R   is catenary and   depth P = depth P∩R , for all   P ∈ Spec R^H .

(A.11.16) R^* satisfies the f.c.c.

(A.11.17) $R(X) = R[X]_{MR[X]}$ satisfies the s.c.c.

(A.11.18) $R[X]$ is catenary.

(A.11.19) $D = R[X]_{(M,X)}$ is catenary.

(A.11.20) $R\langle X\rangle$ is taut-level, where $R\langle X\rangle = R[X]_S$ with $S = R[X] - \cup\{N \in \text{Spec } R[X] \ ; \ MR[X] \subset N\}$.

(A.11.21) For some i $(1 \leqq i \leqq a-1)$, $R[X_1,\ldots,X_i]_{(M,X_1,\ldots,X_i)}$ is an H_j-domain, for $j = i, i + 1,\ldots,a-1$.

(A.11.22) For all y in the quotient field of R such that $(M,y)R[y]$ is proper, $A = R[y]_{(M,y)}$ is catenary and altitude $A = a$.

(A.11.23) Every DVR (V,N) in the quotient field of R such that $R \subseteq V$, $N \cap R = M$, and V is integral over a locality over R is of the first kind.

(A.11.24) For each ideal I of the <u>principal class</u> (that is, I has a basis of height I elements) in R , every prime divisor of $(I^n)_a$, the integral closure in R of I^n , has height = height I .

(A.11.25) For all M-primary ideals $C \subseteq B$ in R such that $e(C) = e(B)$ (with e denoting multiplicity), C is a <u>reduction</u> of B (that is, $CB^n = B^{n+1}$, for all large n) .

(A.11.26) For all ideals I (or, for all M-primary ideals I) in R , all the minimal prime ideals in the associated graded ring of R with respect to I have the same depth.

(A.11.27) For all ideals I (or, for all M-primary ideals I) in R , all maximal relevant ideals in the Rees ring $\mathfrak{R}(R,I)$ of R with respect to I have the same height.

(A.11.28) There exists an M-primary ideal I in R such that I is generated by a system of parameters and height $p'\cap\mathfrak{R}(R,I) = 1$, for all prime divisors p' of $u\mathfrak{R}(R,I)'$.

(A.11.29) There exists an M-primary ideal I in R such that I is generated by a system of parameters and $\mathfrak{R}[1/u]\cap\mathfrak{R}^{(1)} \subseteq \mathfrak{R}'$, where $\mathfrak{R} = \mathfrak{R}(R,I)$.

(A.11.30) For all ideals I in R, $\mathfrak{R}[1/u]\cap\mathfrak{R}^{(1)} \subseteq \mathfrak{R}'$, where $\mathfrak{R}$

113

$= \Re(R,I)$.

(A.11.31) For all $b \in M$, $\mathfrak{L}(R,bR)$ is taut, where $\mathfrak{L}(R,bR)$ is as in (A.4.8).

(A.11.32) $L^{(1)} \subseteq L'$, for all localities L over R .

Proof. (A.11.1) $\Leftrightarrow$ (A.11.2) $\Leftrightarrow$ (A.11.4), by [R-2, Theorem 3.1]; and, (A.11.3) $\Leftrightarrow$ (A.11.4), by [R-5, Theorem 3.3].

(A.11.1) $\Leftrightarrow$ (A.11.5) $\Leftrightarrow$ (A.11.18), by [R-2, Corollary 3.7 and Theorem 3.6]; and, (A.11.1) $\Leftrightarrow$ (A.11.8) $\Leftrightarrow$ (A.11.9), by [R-2, Theorem 3.11].

It is clear that (A.11.1) $\Rightarrow$ (A.11.6), and (A.11.6) $\Rightarrow$ (A.11.1), by [N-6, (34.2)].

(A.11.1) $\Rightarrow$ (A.11.7), by the definition of s.c.c.; and, (A.11.7) $\Rightarrow$ (A.11.12) $\Rightarrow$ (A.11.10) $\Rightarrow$ (A.11.1), by [R-6, (4.3)] and [R-3, Theorem 2.19].

(A.11.1), (A.11.11), and (A.11.13) are equivalent, by [RP, (3.11)] and [R-16, (3.10)], respectively, and (A.11.1) $\Leftrightarrow$ (A.11.14) $\Rightarrow$ (A.11.16) $\Leftrightarrow$ (A.11.19), by [R-3, Theorem 2.21].

(A.11.14) implies R is catenary (since (A.11.14) $\Rightarrow$ (A.11.1)) and height P + depth P = altitude R^H , for all $P \in \text{Spec } R^H$, so (A.11.14) $\Rightarrow$ (A.11.15) (since height P = height $P \cap R$). And, conversely, (A.11.15) implies R^H is taut-level, since height P = height $P \cap R$, so (A.11.15) $\Rightarrow$ (A.11.14), by [McR-2, Proposition 7].

(A.11.1), (A.11.17), and (A.11.20) - (A.11.23) are all equivalent, by, respectively, [RMc, (2.15)], [Mc-2, Corollary 5], [RP, (3.11)] together with [R-15, (2.6)], [RP, (3.16)], and [R-13, (2.11)].

(A.11.4) and (A.11.24) - (A.11.27) are all equivalent, by [R-8, Theorem 2.29] and [R-3, Theorems 3.1 and 3.8]; and, finally, (A.11.4) and (A.11.28) - (A.11.32) are all equivalent, by [Saw, Corollary 4.17 and Theorem 4.16], and [R-9, (3.18.6)], q.e.d.

An example related to (A.11.24) is given in (B.5.6).

APPENDIX B

NOTES ON M. NAGATA'S CHAIN PROBLEM EXAMPLE(S)

In this appendix we first summarize M. Nagata's construction in [N-3, Section 3] (and repeated in [N-6, Example 2, pp. 203-205]) of a family of local domains to show that the answer to the chain problem of prime ideals is no. Then, since these rings have proved to be very useful in answering a number of other questions about saturated chains of prime ideals, we prove a few properties of them, in (B.2), and then list a number of the ways they have previously been used in the literature, in (B.3) - (B.5). We begin with the construction of the rings.

(B.1) CONSTRUCTION. Let K be a field and let $X, Y_1, \ldots, Y_m$ ($m \geq 0$) be indeterminates. Let $z_i = \sum_1^\infty a_{ij} X^j$ ($a_{ij} \in K$ and $i = 1, \ldots, r > 0$) be elements in $K[[X]]$ that are algebraically independent over $K(X)$, let $z_{i1} = z_i$, and, for $j > 1$, let $z_{ij} = z_i - \sum_{k<j} a_{ik} X^k / X^{j-1}$. Let $A = K[X, \{z_{ij}\}, Y_1, \ldots, Y_m]$, $M = (X, Y_1, \ldots, Y_m)A$, $V = A_M$, $N = (X-1, z_1, \ldots, z_r, Y_1, \ldots, Y_m)A$, $W = A_N$, $R' = V \cap W$, and $R = K+J$, where J is the Jacobson radical of R'.

It is proved in [N-3] (and in [N-6]) that R is a local domain with maximal ideal J, R' is the integral closure of R, $R' = A_{A-(M \cup N)}$, R' is a regular domain with exactly two maximal ideals (MR' and NR'), height $MR' = m+1$, height $NR' = r+m+1$ and $R'/MR' = K = R'/NR'$. Also, the multiplicity $e(J)$ of J is one and R is not regular. Finally, R is catenary when and only when $m = 0$.

We now list a few additional facts concerning R and R'. ((B.2.2) - (B.2.5) were essentially shown to hold for all special extension domains, in [R-6, (4.8)].)

(B.2) REMARK. With the above notation, the following statements hold:

(B.2.1) R' is a special extension ring of R .

(B.2.2) $R' = R[a] = R + aR = R + aK$, for all $a \in R'$, $\notin R$.

(B.2.3) For all ideals I in R' , either $R'/I = R/(I \cap R)$ (if $I \not\subseteq J$) or R'/I is a special extension ring of $R/(I \cap R)$ (if $I \subseteq J$) .

(B.2.4) There exists a one-to-one correspondence between the prime ideals $p \neq J$ in R and the prime ideals $P \notin \{MR', NR'\}$ in R' given by $R_p = R'_p$, so R_p is a regular local domain.

(B.2.5) If $p \neq J$ is a prime ideal in R , then pR' is semi-prime. In fact, if q is p-primary and $q^* = qR_p \cap R'$, then $q = qR'$ and either: (a) $qR' = q^*$ (if $q^* \subseteq J$) ; (b) $qR' = q^* \cap MR'$ (if $q^* \subseteq NR'$, $\not\subseteq MR'$) ; or, (c) $qR' = q^* \cap NR'$ (if $q^* \subseteq MR'$, $\not\subseteq NR'$) . There-fore, in cases (b) and (c), q^i is not p-primary, for $i \geq 2$.

(B.2.6) For each nonzero $b \in J$, J is the only imbedded prime divisor of bR . Moreover, if $c \in bR:J$, $\notin bR$, then $bR' = (bR)_a = (b,c)R$ and $bR:cR = J$.

(B.2.7) $R^{(1)} = R'$, if $m > 0$; $R^{(1)} = W$, if $m = 0$.

(B.2.8) If $S = \{Xz_1, \ldots, Xz_r, XY_1, \ldots, XY_m\}$ and if $a_1, \ldots, a_k$ are distinct elements in S , then height $(a_1, \ldots, a_k)R' = 1$ and $(a_1, \ldots, a_k, X-1)R'$ is a height $k+1$ prime ideal.

Proof. (B.2.1) Since $MR' \cap NR' = (MR')(NR')$, it follows that $J = (X^2-X, z_1, \ldots, z_r, Y_1, \ldots, Y_m)R'$, so $(J,X)R' = MR'$ and $(J,X-1)R' = NR'$. Therefore, by definition (1.1.13), it suffices to prove (B.2.2).

(B.2.2) Let $b \in R'$. Then, since $R'/MR' = K = R'/NR'$, there exist $k_1, k_2 \in K$ such that $b-k_1 \in MR'$ and $b-k_2 \in NR'$. Thus, since $X \in MR'$ and $X-1 \in NR'$, $b - k_1 + (k_1-k_2)X = (b-k_1)(1-X) + (b-k_2)X \in J \subseteq R$. Therefore $b \in K + XK$, so $R' = R + XK = R[X]$. Hence, if $a \in R'$, $\notin R$, then $a = r + kX$, for some $r \in R$ and $0 \neq k \in K$, and so $X = (1/k)(a-r) \in R + aK$, and so $R' = R + XK \subseteq R + aK \subseteq R[a] \subseteq R'$.

(B.2.3) follows easily from (B.2.2) and the fact that J is the maximal ideal in R , and (B.2.4) follows from the fact that J is the conductor of R in R' .

(B.2.5) If $q^* \subseteq NR'$, then $q^* \cap MR' = q^* \cap (NR' \cap MR') = q^* \cap J = q^* \cap R$ $= q \subseteq qR' \subseteq q^* \cap J$, so either $q = qR' = q^* \cap MR'$ (if $q^* \subseteq NR, \not\subseteq MR'$) or $q = qR' = q^*$ (if $q^* \subseteq MR' \cap NR' = J$) . (c) is proved in a similar way.

(B.2.6) Let $0 \neq b \in J$. Then $(bR)_a = bR' \cap R$, by [R-8, Lemma 2.3] , and $bX \in bR' \cap R, \not\in bR$, so $bR \subset (bR)_a$. Now, if $p \neq J$ is a prime divisor of bR and P is the corresponding prime ideal in R' (see (B.2.4)), then $bR_p \subseteq (bR)_a R_p \subseteq (bR')R'_P = bR'_P = bR_p$, so from a primary decomposition of bR it follows that J is a prime divisor of bR (since $bR \subset (bR)_a$) . Also, pR_p is a prime divisor of bR_p and R_p is a regular local ring, by (B.2.4), so J is the only imbedded prime divisor of bR . Let $c \in bR:J, \not\in bR$. Then $cJ \subseteq bR$, so it follows that $J = bR:cR$. Also, $c/b \in R'$, by [N-6, (12.3)], so $c \in bR' \cap R = (bR)_a$. Finally, if $d \in (bR)_a, \not\in bR$, then $d/b \in R'$, $\not\in R$, so $R' = R + (d/b)R$, by (B.2.2). Therefore, for each $r' \in R'$, there exist $r,s \in R$ such that $r' = r + s(d/b)$, and so $br' \in (b,d)R$ Hence $(bR)_a \subseteq bR' \subseteq (b,d)R \subseteq (bR)_a$, so $bR' = (bR)_a = (b,d)R$, and so it follows that $bR' = (bR)_a = (b,c)R$.

(B.2.7) If $m > 0$, then height $MR' > 1$, so, by (B.2.4), there exists a one-to-one correspondence between the height one prime ideals p in R and p' in R' such that $R_p = R'_{p'}$, and so $R^{(1)} = R'^{(1)}$ $= R'$. If $m = 0$, then height $MR' = 1$, so there exists a one-to-one correspondence between the height one prime ideals p in R and the height one prime ideals p' in R' that are contained in NR' such that $R_p = R'_{p'}$, and so $R^{(1)} = (R'_{NR'})^{(1)} = R'_{NR'} = W$.

(B.2.8) follows from the facts: $W = R'_{NR'}$ is a regular local domain whose maximal ideal is generated by $X-1, z_1, \ldots, z_r, Y_1, \ldots, Y_m$; X is a unit in W ; and, NR' is the only maximal ideal in R' that contains X-1 , q.e.d.

Quite a few additional interesting and useful facts concerning R
and R' are proved in [D].

As has already been mentioned, this family of local rings has been
used in a number of papers in the literature to answer certain questions concerning saturated chains of prime ideals in a Noetherian ring.
Some of these applications are listed in (B.3) - (B.5), and among these
are results that show: the existence of rings that have certain properties; that certain hypotheses for a known result cannot be weakened;
and, that the converse of certain known results does not hold. We re-
strict attention to the case: $m = 0$, in (B.3); $m \geqq 0$, in (B.4); and,
$m > 0$, in (B.5). A few of the results in (B.3) - (B.5) are new, so a
proof of these will be given. Also, a brief explanation as to why each
of the results is of interest and/or importance will be given following
each remark. Finally, as in Appendix A, there are quite a few other
applications of these rings in the literature that are not included in
(B.3) - (B.5), but the lists do show the variety of uses that have been
made of these rings.

(B.3) REMARK. Let the notation be as in (B.1), let $D = R[T]_{(J,T)}$,
and let $m = 0$. Then the following statements hold:

(B.3.1) [McR-2, p. 74]. R is taut-level, but R' is only taut,
not taut-level.

(B.3.2) R satisfies the o.h.c.c., but not the s.c.c.

(B.3.3) For all nonzero $p \in$ Spec R , R/p satisfies the s.c.c.,
and for all non-maximal $P \in$ Spec R , R_p satisfies the s.c.c., but R
does not satisfy the s.c.c.

(B.3.4) [R-2, Remark 3.9]. R' satisfies the c.c., and R does
not satisfy the c.c.

(B.3.5) [R-12, (4.1.7)]. With $\mathcal{S}$ as in (9.2.4), $D \in \mathcal{S}$, but

D/TD and D_{JD} are not in $\mathcal{S}$.

(B.3.6) [R-11, (2.16)]. For all nonzero b in J , D[1/b] satisfires the c.c., but D[1/T] does not satisfy the c.c.

(B.3.7) [R-15, comment preceding (3.2)]. If r = 2 , then D
is an H_2-domain, but is not an H_1-domain.

(B.3.8) [HMc, Example 3.1]. If r = 1 , then H = {P $\in$ Spec D ;
D_P is not catenary} is not closed in the Zariski topology.

(B.3.9) [HMc, Example 3.2]. If r = 1 , then I = {Q $\in$ Spec D ;
height Q = 1 = depth Q} is an infinite set, but (J,T)D $\neq$ $\cup$ {Q ; Q $\in$ I}.

(B.3.10) [R-9, (3.18.5)]. If r = 1 , then, for all nonzero b
in J , $\mathcal{L}$ = $\mathcal{L}$(R,bR) is homogeneously taut-level, but $\mathcal{L}$ is not an
H-domain.

(B.3.11) [Fu-1, Lemma 6]. If r = 1 and the characteristic of
K is zero, then B = R$\cap$K[X,z_1,1/X] is a Noetherian Hilbert domain
that satisfies the f.c.c. but not the s.c.c.

(B.3.12) [R-18, (3.2) and (3.3)]. If r = 1 and B is as in
(B.3.11), then B[T] is a Noetherian Hilbert domain that satisfies the
o.h.c.c., but the following statements do not hold: B[T] is catenary;
B[T]' is taut; B[T]' satisfies the o.h.c.c.; and, B[T]$_P$ satisfies
the o.h.c.c., for all P $\in$ Spec B[T] . Moreover, B[T_1,T_2] satisfies
the o.h.c.c., but B[T_1,T_2]/P does not satisfy the o.h.c.c., for some
P $\in$ Spec B[T_1,T_2] .

Proof. (B.3.2) R is taut (since R is catenary (since m = 0)) ,
and R' satisfies the c.c., since R' is a regular domain, so R
satisfies the o.h.c.c. R does not satisfy the s.c.c., since R' is
not level.

(B.3.3) If (0) $\neq$ p $\in$ Spec R and p' $\in$ Spec R' lies over p ,
then p' $\not\subseteq$ J (since m = 0) , so R'/p' = R/p , by (B.2.3). Now
R'/p' satisfies the c.c., by (1.3.1), so R/p satisfies the s.c.c.
Also, if J $\neq$ P $\in$ Spec R , then R_P = R'$_{(R-P)}$ is a regular local ring,

by (B.2.4), so R_p satisfies the s.c.c. Finally, R does not satisfy the s.c.c., by (B.3.2), q.e.d.

(B.3.1) shows that "taut-level" is not inherited by integral extension domains, whereas "taut" is, by [McR-2, Proposition 12].

(B.3.2) (and (1.1.7) together with (A.11.1) ⇒ (A.11.7)) shows that "o.h.c.c." is a weaker condition on a local domain than "s.c.c."

If (L,P) is a local domain that satisfies the s.c.c. and Q ∈ Spec L is such that (0) ≠ Q ≠ P , then (1.3.1) shows that L/Q and L_Q satisfy the s.c.c. (B.3.3) shows that the converse does not hold.

(B.3.4) shows that the c.c. part of [N-6, (34.2)] is false.

By (A.6.1) implies (A.6.4) and (A.6.6), the class of GB-rings is closed under passage to factor rings and quotient rings. (B.3.5) shows that this does not hold for the closely related class of rings $.

(A.9.8) ⇒ (A.9.7) shows that if a local domain (L,P) is such that L[1/b] satisfies the s.c.c., for some 0 ≠ b ∈ P , then L[1/b] satisfies the s.c.c., for all 0 ≠ b ∈ P . (B.3.6) shows that this does not hold for the c.c.

In [R-15, (3.1)] it is shown that if (L,P) is a local domain and k is a positive integer such that $D_k = L[T_1,\dots,T_k]_{(P,T_1,\dots,T_k)}$ is an H_i-domain, for some i ≤ k , then D_k is an H_j-domain, for j = 0,1,...,i . (B.3.7) shows that this does not hold for i such that k < i < altitude D_k - 1 .

It is always of interest to know whether a given subset of the prime spectrum of a ring is or is not closed, and (B.3.8) answers this for H .

(B.3.9) shows that a certain generalization of the Principal Ideal Theorem does not hold. That is, if it were always true that a prime ideal P of little height two in a Noetherian domain A was such that P = ∪{p ∈ Spec A ; (0) ⊂ p ⊂ P is saturated} , then an interesting generalization of the Principal Ideal Theorem could be shown to hold.

For (B.3.10), a homogeneous ideal in $\mathcal{L}$ is defined to be an ideal of the form $H\mathcal{L}$, where H is a homogeneous ideal in $\Re(R.bR)$. If it were true that homogeneous taut-level local domains of the form $\mathcal{L}(L,bL)$ were always taut, (where (L,P) is a local domain and $0 \neq b \in P$), then it could be shown that the Catenary Chain Conjecture holds, by [R-9, (2.10.1), (3.18.1), (3.18.6), and (3.22)].

(B.3.11) answers a question I asked in [R-11, (2.20)].

It is known [R-18, Section 2] that if S is a semi-local ring that satisfies the o.h.c.c., then: S is catenary; S' is taut; S' satisfies the o.h.c.c.; and, for all $P \in \text{Spec } S$, S_p and S/P satisfy the o.h.c.c. (B.3.12) shows that these facts cannot be extended to all Noetherian rings that satisfy the o.h.c.c.

(B.4) REMARK. Let the notation be as in (B.1), let $D = R[T]_{(M,T)}$, and let $m \geqq 0$. Then the following statements hold:

(B.4.1) [P, p. 8]. R is an H_i-domain if, and only if, either $i = 0$ or $i \geqq m+1$.

(B.4.2) R is a C_i-domain if, and only if, $i \geqq m+1$.

(B.4.3) [P. p. 68]. R is a D_i-domain if, and only if, either $i = 0$ or $i \geqq m+1$.

(B.4.4) R_p satisfies the s.c.c., for all non-maximal $P \in \text{Spec } R$, but R is not catenary if $m > 0$.

(B.4.5) [R-19, final paragraph]. R' is a GB-domain, but R is not a GB-domain.

(B.4.6) [R-12, p. 124]. If there exists a mcpil n in some integral extension domain of $R(T) = R[T]_{JR[T]}$, then there may not exist a mcpil n in $R(T)$.

(B.4.7) [R-20, (2.8.1)]. Let $S = R'[T]_{(R[T] - (J,T))}$ and let $L = D+J'$, where J' is the Jacobson radical of S. Then $D \in \mathcal{S}$, where $\mathcal{S}$ is as in (9.2.4), L is a finite local integral extension domain of D , and there exists $P \in \text{Spec } L$ such that height $P <$ height $P \cap D$.

(B.4.8) [R-20, (2.10)]. With D , S , and L as in (B.4.7), there exists a mcpil $m+1$ in L that does not contract to a maximal chain of prime ideals in D .

Proof. (B.4.2) Clearly R is a C_{r+m+1}-domain, and R is not a C_i-domain for $i < m+1$, by (B.4.1). Finally, let $i \in \{m+1,\ldots,r+m+1\}$, let $p \in \text{Spec } R$ such that height $p = i$, and let $P \in \text{Spec } R'$ such that $P \cap R = p$. Then height $P = $ height p (by (B.2.4)) and $P \not\subseteq J$ (since $m+1 \leq i$) , so $R/p = R'/P$ satisfies the s.c.c., by the proof of (B.3.3). Thus $(R/p)'$ is level, so R is a C_i-domain, by (B.4.1). (B.4.4) follows from (B.2.4), q.e.d.

After defining an H_i-ring (respectively, C_i-ring, D_i-ring), it is of interest to know if there exists a non-catenary local domain L that is an H_j-ring (respectively, C_i-ring, D_i-ring) for some i $(0 < i < \text{altitude } L)$. (B.4.1) - (B.4.3) show that such rings do indeed exist.

By [R-4, Corollary 3.13], a local domain (L,P) is catenary if, and only if, for all nonmaximal $p \in \text{Spec } L$, L_p satisfies the s.c.c. and height p + depth p = altitude L . (B.4.4) (for $m > 0$) shows that the condition "height p + depth p = altitude L" is necessary.

By (A.6.1) $\Rightarrow$ (A.6.2), an integral extension domain of a GB-local domain is a GB-domain. (B.4.5) shows that the converse of this does not hold.

By [RMc, (2.14)], if (L,P) is a local domain, then there exists a mcpil n in some integral extension domain of $C = L[T]_{(P,T)}$ if, and only if, there exists a mcpil n in C . (B.4.5) shows that this does not continue to hold for local domains of the form $L(T)$.

(B.4.7) and (B.4.8) answer questions asked in [R-10, (3.15)] and they show that two bad things which were shown to be possible by Nagata's examples can also be shown to be possible in a local (rather than a semi-local) integral extension domain.

(B.5)   REMARK.   Let the notation be as in (B.1), let $D = R[T]_{(J,T)}$, and let $m > 0$. Then the following statements hold:

(B.5.1)   [R-2, Remark 3.9]. R' is catenary and satisfies the c.c., but R is neither catenary nor satisfies the c.c.

(B.5.2)   [R-3, Remark 5.12]. $R^{(1)}$ is a finite R-algebra and R is not quasi-unmixed.

(B.5.3)   [R-3, p. 127].   There exist analytically independent elements b,c in R such that $JR[c/b]$ is a depth one prime ideal and height $JR[c/b] <$ height $J - 1$.

(B.5.4)   If $r = m = 1$, then $R(c/b)$ is catenary, for all ana- lytically independent elements b,c in R , but R is not catenary.

(B.5.5)   [R-8, Example 2.28(a)].   With $q = (X,Y_1,\dots,Y_{m-1})R'\cap R$, $p = (X-1)R'\cap R$, $Q = R[T_1,T_2]_{(J,T_1,T_2)}$, and $L = Q/(p\cap q)Q$, L^* is a complete local ring that has a regular element c and a minimal prime ideal z such that $(z,c)L^*$ is a non-maximal prime ideal of height > 1 .

(B.5.6)   [R-8, Example 2.28(b)].   If $m = r = 1$, if L is as in (B.5.5), and if P is the maximal ideal in L , then $I = (Y_1,z_1)L$ is a height two ideal of the principal class in L and $(I^i)_a:P = (I^i)_a$, for all $i \geq 1$, and L is not quasi-unmixed.

(B.5.7)   [R-11, (2.22)]. $D[1/T]$ is a Noetherian Hilbert domain such that there exists a maximal ideal P in $D[1/T]'$ such that height P < height $P\cap D[1/T]$.

(B.5.8)   If $m = 1$, then there exists a prime ideal P in D such that height $P = 2$, depth $P = 1$, and there exist infinitely many $p^* \in$ Spec D such that $p^* \subset P$ and depth $p^* >$ depth $P + 1$.

(B.5.9)   [R-20, (2.6)]. $q = ((X-1,Y_1,\dots,Y_{m-1})R'\cap R,T-Y_m)D \in$ Spec D is such that $Q = (q,T)D \in \text{Spec } D$ and $m+r+2 = \text{height } Q > \text{height } q + 1$ $= m+2$.

Proof. (B.5.8) Since $m = 1$, there exists a height two maximal ideal in R' , so there exist infinitely many height one depth one prime ideals in R' , and so there exists a height one depth one prime ideal p in R , by [Mc-1, Theorem 7]. Let q be a height one prime ideal in R such that depth $q = \text{altitude } R - 1$, so $r+1 = \text{depth } q$ > 1 . Let $P = (p,T)D$ and $Q = (q,T)D$, so height $P = 2 = \text{height } Q$, depth $P = 1$, and depth $Q = r+1$. Then, by [Mc-5, Theorem 3], there exist infinitely many height one prime ideals p^* in D such that $p^* \subset P \cap Q$, so depth $p^* = \text{depth } Q + 1 > 2 = \text{depth } P + 1$, q.e.d.

(B.5.1) shows that the c.c. part of [N-6, (34.2)] is false, and it shows that the analogous statement for "catenary" in place of "c.c." is also false.

It is known [R-3, Lemma 5.11] that if R is an unmixed local domain, then $R^{(1)}$ is a finite R-algebra. (B.5.2) shows that the converse of this does not hold. (As mentioned in (15.6.3), it is an open problem if integrally closed local domains are unmixed - and for such R , $R^{(1)} = R$.)

As noted in $(4.1.1) \Leftrightarrow (4.1.4)$, it is an open problem if there exists a Henselian local domain (L,P) that is not an H-domain ; that is, by $(A.3.1) \Leftrightarrow (A.3.7)$, does there exist such (L,P) such that there are analytically independent elements b,c in L such that height $PL[c/b] < \text{height } P - 1$. (B.5.3) shows that this can happen for non-Henselian local domains.

It is known [R-4, Theorem 4.12] that a local domain (L,P) is catenary if, and only if, for all analytically independent elements b,c in L , $L(c/b)$ is catenary and altitude $L(c/b) = \text{altitude } L - 1$. (B.5.4) shows that the hypothesis "altitude $L(c/b) = \text{altitude } L - 1$"

is necessary.

It is known [R-8, Lemma 2.22] that if b is a regular element in B', where B is a Noetherian ring, and if z is a minimal prime ideal in B' such that $(z,b)B' \neq B'$, then height $(z,b)B' = 1$. (B.5.5) shows that the hypothesis about being integrally closed is necessary.

It is known [R-8, Corollary 2.31] that a local ring (L,P) is quasi-unmixed if and only if for all ideals I of the principal class in L such that height I = altitude $L - 1$, $(I^i)_a : P = (I^i)_a$, for all large i. (B.5.6) shows that the hypothesis "for all" cannot be replaced by "there exists."

(B.5.7) shows that there are Noetherian Hilbert domains that do not satisfy the c.c. Fujita's example (B.3.11) shows the stronger result that there are such domains that satisfy the f.c.c.

(B.5.8) shows that [Mc-1, Theorem 1] cannot be "inverted."

Finally, if B is a Noetherian ring and $P \in \text{Spec } B$, then height $(P,T)B[T]$ = height $PB[T] + 1$. (B.5.9) shows that this does not continue to hold for non-extended prime ideals in $B[T]$.

BIBLIOGRAPHY

[B]   J. Brewer, The ideal transform and overrings of an integral
         domain, Math. Z. 107(1968), 301-306.

[Bro]   M. Brodmann,"Über die minimale dimension der assoziierten
         primideal der komplettion eines lokalen integritaetsbereiches,"
         Ph.D. Dissertation, University of Basel, Switzerland, 1974.

[C-1]   I. S. Cohen, On the structure and ideal theory of complete
         local rings, Trans. Amer. Math. Soc. 59(1946), 54-106.

[C-2]   _____, Lengths of prime ideal chains, Amer. J. Math.
         76(1954), 654-668.

[D]   L. Déchene,"Adjacent integral extension domains,"Ph.D. Dis-
         sertation, University of California, Riverside, in prepara-
         tion.

[FR]   D. Ferrand and M. Raynaud, Fibres formelles d'un anneau lo-
         cal Noethérien, Ann. Sci. Ecole Norm. Sup. 3(1970), 295-311.

[F-M]   M. Flexor-Mangeney, Étude de l'assassin du complété d'un
         anneau local Noethérien, Bull. Soc. Math. France 98(1970),
         117-125.

[Fu-1]   K. Fujita, Some counterexamples related to prime chains of
         integral domains, Hiroshima Math. J. 5(1975), 473-485.

[Fu-2]   _____, Three dimensional unique factorization domain
         which is not catenarian, 7 page preprint.

[G-1]   A. Grothendieck, "Éléments de Géométrie Algébrique," IV
         (Premiere Partie), Inst. Hautes Etudes Sci. Publ. Math.
         Presses Universitaires de France, Paris, France, 1964.

[G-2]   _____, "Éléments de Géométrie Algébrique," IV (Seconde
         Partie), Inst. Hautes Etudes Sci. Publ. Math. Presses Uni-
         versitaires de France, Paris, France, 1965.

[G-3]   _____, "Éléments de Géométrie Algébrique," IV (Quar-
         trieme Partie), Inst. Hautes Etudes Sci. Publ. Math. Presses
         Universitaires de France, Paris, France, 1967.

[H]   R. C. Heitmann, Prime spectra in Noetherian rings, 8 page
         preprint.

[Hoc]   M. Hochster, "Topics in the Homological Theory of Modules
         over Commutative Rings," Conference Board of the Mathe-
         matical Sciences Regional Conference Series in Mathematics,
         No. 24, Amer. Math. Soc., Providence, RI, 1975.

[HMc]   E. G. Houston and S. McAdam, Chains of primes in Noetherian
         rings, Indiana Univ. Math. J. 24(1975), 741-753.

[Hou-1] _____, Localizations of H_i and D_i rings, Canadian
         J. Math., forthcoming.

[Hou-2] _____, Chains of primes in $R\langle X\rangle$, 21 page preprint.

[K]   I. Kaplansky, Adjacent prime ideals, J. Algebra 20(1972),
         94-97.

[Kr]   W. Krull, Zum Dimensionsbegriff der Idealtheorie (Beiträge
         zur Arithmetik kommutativer Integritätsbereiche, III), Math.
         Z. 42(1937), 745-766.

[L]   W. J. Lewis, The spectrum of a ring as a partially ordered set, J. Algebra 25(1973), 419-434.

[M]   H. Matsumura, "Commutative Algebra," W. A. Benjamin, Inc.,
          New York, NY, 1970.

[Mc-1]   S. McAdam, Saturated chains in Noetherian rings, Indiana Univ.
          Math. J. 23(1974), 719-728.

[Mc-2]   _____, On taut-level R⟨X⟩, Duke Math. J. 42(1975), 637-
          644.

[Mc-3]   _____, A Noetherian example, Comm. Algebra 4(1976), 245-
          247.

[Mc-4]   _____, Primes of little height 2 in polynomial rings,
          13 page preprint.

[Mc-5]   _____, Intersections of height 2 primes, 15 page pre-
          print.

[McR-1]   _____ and L. J. Ratliff, Jr., On prime ideals p ⊂ P∩Q,
          Duke Math. J. 42(1975), 313-319.

[McR-2]   _____ and _____, Semi-local taut rings, Indiana Univ.
          Math. J. 26(1977), 73-79.

[N-1]   M. Nagata, On the theory of Henselian rings, II, Nagoya Math.
          J. 7(1954), 1-19.

[N-2]   _____, An example of normal ring which is analytically
          ramified, Nagoya Math. J. 9(1955), 111-113.

[N-3]   _____, On the chain problem of prime ideals, Nagoya Math.
          J. 10(1956), 51-64.

[N-4]   _____, An example of a normal local ring which is analy-
          tically reducible, Mem. Coll. Sci., Univ. Kyoto 31(1958),
          83-85.

[N-5]   _____, Note on a chain condition for prime ideals, Mem.
          Coll. Sci., Univ. Kyoto 32(1959-1960), 85-90.

[N-6]   _____, "Local Rings," Interscience Tracts 13, Inter-
          science, New York, NY, 1962.

[Ni]   M. Nishi, On the dimension of local rings, Mem. Coll. Sci.,
          Univ. Kyoto 29(1955), 7-9.

[P]   M. E. Pettit, Jr., "Properties of H_i-rings," Ph.D. Disserta-
          tion, University of California, Riverside, 1973.

[R-1]   L. J. Ratliff, Jr., On quasi-unmixed semi-local rings and the
          altitude formula, Amer. J. Math. 87(1965), 278-284.

[R-2]   _____, On quasi-unmixed local domains, the altitude for-
          mula, and the chain condition for prime ideals (I), Amer. J.
          Math. 91(1969), 508-528.

[R-3]   _____, On quasi-unmixed local domains, the altitude for-
          mula, and the chain condition for prime ideals (II), Amer. J.
          Math. 92(1970), 99-144.

[R-4]   _____, Characterizations of catenary rings, Amer. J.
          Math. 93(1971), 1070-1108.

[R-5]   _____, Catenary rings and the altitude formula, Amer.
          J. Math. 94(1972), 458-466.

[R-6]   _____, Chain conjectures and H-domains, pp. 222-238,
          Lecture Notes in Mathematics, No. 311; "Conference on Commu-
          tative Algebra," Springer-Verlag, New York, NY, 1973.

[R-7] _____, Three theorems on imbedded prime divisors of principal ideals, Pacific J. Math. 49(1973), 199-210.

[R-8] _____, Locally quasi-unmixed Noetherian rings and ideals of the principal class, Pacific J. Math. 52(1974), 185-205.

[R-9] _____, On Rees localities and H_i-local rings, Pacific J. Math. 60(1975), 169-194.

[R-10] _____, Four notes on saturated chains of prime ideals, J. Algebra 39(1976), 75-93.

[R-11] _____, Hilbert rings and the chain condition for prime ideals, J. Reine Angew. Math. 283/284(1976), 154-163.

[R-12] _____, Maximal chains of prime ideals in integral extension domains, II, Trans. Amer. Math. Soc. 224(1976), 117-141.

[R-13] _____, The altitude formula and DVR's, Pacific J. Math., 67(1976), 509-523.

[R-14] _____, Characterizations of taut semi-local rings, Ann. Math. Pura Appl. 112(1977), 151-192.

[R-15] _____, Polynomial rings and H_i-local rings, Pacific J. Math., forthcoming.

[R-16] _____, Going-between rings and contractions of saturated chains of prime ideals, Rocky Mountain J. Math., forthcoming.

[R-17] _____, H-semi-local domains and altitude R[c/b], Proc. Amer. Math. Soc., forthcoming.

[R-18] _____, Notes on a new chain condition for prime ideals, J. Pure Appl. Algebra, forthcoming.

[R-19] _____, A(X) and GB-Noetherian rings, Rocky Mountain J. Math., forthcoming.

[R-20] _____, Notes on local integral extension domains, Canadian J. Math., forthcoming.

[R-21] _____, Valuation rings and the (catenary) chain conjectures, 92 page preprint.

[R-22] _____, Notes on saturated chains of prime ideals, in preparation.

[RMc] _____ and S. McAdam, Maximal chains of prime ideals in integral extension domains, I, Trans. Amer. Math. Soc. 224 (1976), 103-116.

[RP] _____ and M. E. Pettit, Jr., Characterizations of H_i-local rings and of C_i-local rings, Amer. J. Math., forthcoming.

[Re] D. Rees, A-transforms of local rings and a theorem on multiplicities of ideals, Proc. Cambridge Philos. Soc. 57(1961), 8-17.

[Ru] D. Rush, Some applications of Griffith's basic submodules, J. Pure Appl. Algebra, forthcoming.

[S] J. D. Sally, Failure of the saturated chain condition in an integrally closed domain, Notices Amer. Math. Soc. 17(1970), 70T - A72, p. 560.

[Saw] P. G. Sawtelle, "Characterizations of unmixed and quasi-unmixed local domains," Ph.D. Dissertation, University of California, Riverside, 1971.

[Sei]   A. Seidenberg, A note on the dimension theory of rings,
          Pacific J. Math. 3(1953), 505-512.

[Sey]   H. Seydi, Anneaux Henséliens et conditions de chaînes, III,
          Bull. Soc. Math. France 98(1970), 329-336.

[Y]   M. Yoshida, A theorem on Zariski rings, Canad. J. Math. 8
          (1956), 3-4.

[Z]   O. Zariski, Analytical irreducibility of normal varieties,
          Ann. of Math. 49(1948), 352-361.

[ZS-1]   _____ and P. Samuel, "Commutative Algebra," Vol. I, Van
          Nostrand, New York, NY, 1958.

[ZS-2]   _____ and _____, "Commutative Algebra," Vol. II,
          Van Nostrand, New York, NY, 1960.

SUPPLEMENTAL BIBLIOGRAPHY

[1]   J. T. Arnold, On the dimension theory of overrings of an
          integral domain, Trans. Amer. Math. Soc. 138(1969), 313-326.

[2]   _____ and R. Gilmer, Dimension sequences for commutative
          rings, Bull. Amer. Math. Soc. 79(1973), 407-409.

[3]   _____ and _____, The dimension sequence of a com-
          mutative ring, Amer. J. Math. 96(1974), 385-408.

[4]   _____ and _____, Dimension theory of commutative
          rings without identity, J. Pure Appl. Algebra 5(1974), 209-
          231.

[5]   _____ and _____, The dimension theory of commuta-
          tive semigroup rings, Houston J. Math. 2(1976), 299-313.

[6]   M. Auslander and A. Rosenberg, Dimension of ideals in poly-
          nomial rings, Canad. J. Math. 10(1958), 287-293.

[7]   E. Bastida and R. Gilmer, Overrings and divisorial ideals of
          rings of the form D+M, Mich. Math. J. 20(1973), 79-95.

[8]   M. Brodmann, Ueber die minimale Dimension der assoziierten
          Primideals der Komplettion eines lokalen Integritätsbereiches,
          Comment. Math. Helv. 50(1975), 219-232.

[9]   _____, Dimension des associés du complete d'un anneau
          integre et local, 25 page preprint.

[10]   _____, A "macaulayfication" of unmixed domains, 23 page
          preprint.

[11]   _____, On some results of L. J. Ratliff, 26 page pre-
          print.

[12]   E. D. Davis, Ideals of the principal class, R-sequences and
          a certain monoidal transformation, Pacific J. Math. 20(1967),
          197-205.

[13]   _____, Further remarks on ideals of the principal class,
          Pacific J. Math. 27(1968), 49-51.

[14]   _____ and S. McAdam, Prime divisors and saturated chains,
          19 page preprint.

[15]   J. Dieudonne, "Topics in Local Algebra," Notre Dame Math. Lec-
          tures No. 10, Univ. of Notre Dame Press, Notre Dame, 1967.

[16]   M. Flexor-Mangeney, Étude de certains éclatements, **Bull. Soc.**
        Math. France 100(1972), 229-239.

[17]   _____, Une propriété des anneaux unibranches, **Bull. Sci.**
        Math. 96(1972), 169-175.

[18]   K. Fujita, Infinite dimensional Noetherian Hilbert domains,
        Hiroshima Math. J. 5(1975), 181-185.

[19]   R. Gilmer, Dimension sequences of commutative rings, pp. 31-
        36, "Ring Theory" (Proc. Conf. Univ. Oklahoma, Norman, Okla.,
        1973), Lecture Notes in Pure and Applied Math., Vol. 7,
        Dekker, New York, NY, 1974.

[20]   _____, On polynomial and power series rings over a commu-
        tative ring, **Rocky Mountain J. Math.** 5(1975), 151-175.

[21]   M. Hochster and L. J. Ratliff, Jr., Five theorems on Macaulay
        rings, **Pacific J. Math.** 44(1973), 147-172.

[22]   P. Jaffard, "Théorie de la Dimension dans les Anneaux de
        Polynomes," Mémor. Sci. Math., fasc. 146, Gauthier-Villars,
        Paris, France, 1960.

[23]   I. Kaplansky, "Commutative Rings," Allyn and Bacon, Boston,
        Mass., 1970.

[24]   W. Krull, Primidealketten in allgemeinen Ringbereichen, **S.-B.**
        Heidelberger Akad. Wiss. Math.-Natur. Kl. 7(1928).

[25]   W. Krull, "Idealtheorie," Ergebnisse der Mathematik und Ihrer
        Grenzgebiete, Vol. 4, No. 3, Julius Springer, Berlin, Germany,
        1935.

[26]   _____, Eine Ergänzung von Beitrag III (Beiträge zur
        Arithmetic Kommutativer Integritätsbereiche, IIIa), **Math. Z.**
        43(1938), 767.

[27]   _____, Dimensionstheorie in Stellenringen, **J. Reine**
        Angew. Math. 179(1938), 204-226.

[28]   _____, Jacobsonsche Ringe, Hilbertscher Nullstellensatz,
        Math. Z. 54(1951), 354-387.

[29]   D. C. Lantz, Preservation of local properties and chain con-
        ditions in commutative group rings, **Pacific J. Math.** 63(1976),
        193-199.

[30]   S. McAdam, 1-Going down, **J. London Math. Soc.** 8(1974), 674-
        680.

[31]   _____ and E. G. Houston, Rank in Noetherian rings, **J.**
        Algebra, forthcoming.

[32]   _____, Unmixed 2-dimensional local domains, **Pacific J.**
        Math., forthcoming.

[33]   M. Nagata, Some remarks on the 14th problem of Hilbert, **J.**
        Math. Kyoto Univ. 5(1965-66), 61-66.

[34]   _____, Finitely generated rings over a valuation ring,
        J. Math. Kyoto Univ. 5(1965-66), 163-169.

[35]   D. G. Northcott, On the notion of a form ideal, **Quarterly J.**
        Math. 4(1953), 221-229.

[36]   T. Parker, A number theoretic characterization of dimension
        sequences, **Amer. J. Math.** 97(1975), 308-311.

[37]   L. J. Ratliff, Jr., A characterization of analytically

unramified semi-local rings and applications, Pacific J. Math. 27(1968), 127-143.

[38] L. J. Ratliff, Jr. and S. McAdam, Polynomial rings and H_i-local rings, Houston J. Math. 1(1975), 101-120.

[39] H. Sato, On splitting of valuations in extensions of local domains, J. Sci. Hiroshima Univ. 21(1957), 69-75.

[40] P. G. Sawtelle, Quasi-unmixed local rings and quasi-subspaces, Proc. Amer. Math. Soc. 38(1973), 59-64.

[41] _____, Quasi-unmixedness and integral closure of Rees rings, Proc. Amer. Math. Soc. 56(1976), 95-98.

[42] U. Schweizer, Fortsetzung von Spezialisierungen: ein ideal-theoretischer Zugang, Comment. Math. Helv. 49(1974), 245-250.

[43] A. Seidenberg, On the dimension theory of rings (II), Pacific J. Math. 4(1954), 603-614.

[44] H. Seydi, Anneaux henséliens et conditions de chaînes: La formule des dimensions, C.R. Acad. Sci. Paris Ser. A-B 270 (1970), 696-698.

[45] _____, Anneaux henséliens et conditions de chaînes, C. R. Acad. Sci. Paris Ser. A-B 271(1970), 120-121.

[46] _____, Anneaux henséliens et conditions de chaînes, Bull. Soc. Math. France 98(1970), 9-31.

[47] _____, Sur une note d'Ernst Kunz, C. R. Acad. Sci. Paris Ser. A-B 274(1972), 714-716.

[48] _____, Sur la théorie des anneaux excellents en charac-teristique p, Bull. Sci. Math. 2nd Series 96(1972), 193-198.

[49] _____, La théorie des anneaux japonais, Colloque d'algebre commutative (Rennes, 1972), Exp. No. 12, 82 pp. Publ. Sém. Math. Univ. Rennes, Année, 1972, Univ. Rennes, Rennes, 1972.

[50] _____, Sur le probleme des chaînes d'ideaux premiers dans les anneaux Noetheriens, 8 page preprint.

TABLE OF NOTATION

INDEX

Vol. 489: J. Bair and R. Fourneau, Etude Géométrique des Espaces Vectoriels. Une Introduction. VII, 185 pages. 1975.

Vol. 490: The Geometry of Metric and Linear Spaces. Proceedings 1974. Edited by L. M. Kelly. X, 244 pages. 1975.

Vol. 491: K. A. Broughan, Invariants for Real-Generated Uniform Topological and Algebraic Categories. X, 197 pages. 1975.

Vol. 492: Infinitary Logic: In Memoriam Carol Karp. Edited by D. W. Kueker. VI, 206 pages. 1975.

Vol. 493: F. W. Kamber and P. Tondeur, Foliated Bundles and Characteristic Classes. XIII, 208 pages. 1975.

Vol. 494: A Cornea and G. Licea. Order and Potential Resolvent Families of Kernels. IV, 154 pages. 1975.

Vol. 495: A. Kerber, Representations of Permutation Groups II. V, 175 pages. 1975.

Vol. 496: L. H. Hodgkin and V. P. Snaith, Topics in K-Theory. Two Independent Contributions. III, 294 pages. 1975.

Vol. 497: Analyse Harmonique sur les Groupes de Lie. Proceedings 1973-75. Edité par P. Eymard et al. VI, 710 pages. 1975.

Vol. 498: Model Theory and Algebra. A Memorial Tribute to Abraham Robinson. Edited by D. H. Saracino and V. B. Weispfenning. X, 463 pages. 1975.

Vol. 499: Logic Conference, Kiel 1974. Proceedings. Edited by G. H. Müller, A. Oberschelp, and K. Potthoff. V, 651 pages 1975.

Vol. 500: Proof Theory Symposion, Kiel 1974. Proceedings. Edited by J. Diller and G. H. Müller. VIII, 383 pages. 1975.

Vol. 501: Spline Functions, Karlsruhe 1975. Proceedings. Edited by K. Böhmer, G. Meinardus, and W. Schempp. VI, 421 pages. 1976.

Vol. 502: János Galambos, Representations of Real Numbers by Infinite Series. VI, 146 pages. 1976.

Vol. 503: Applications of Methods of Functional Analysis to Problems in Mechanics. Proceedings 1975. Edited by P. Germain and B. Nayroles. XIX, 531 pages. 1976.

Vol. 504: S. Lang and H. F. Trotter, Frobenius Distributions in GL$_2$-Extensions. III, 274 pages. 1976.

Vol. 505: Advances in Complex Function Theory. Proceedings 1973/74. Edited by W. E. Kirwan and L. Zalcman. VIII, 203 pages. 1976.

Vol. 506: Numerical Analysis, Dundee 1975. Proceedings. Edited by G. A. Watson. X, 201 pages. 1976.

Vol. 507: M. C. Reed, Abstract Non-Linear Wave Equations. VI, 128 pages. 1976.

Vol. 508: E. Seneta, Regularly Varying Functions. V, 112 pages. 1976.

Vol. 509: D. E. Blair, Contact Manifolds in Riemannian Geometry. VI, 146 pages. 1976.

Vol. 510: V. Poènaru, Singularités C$^\infty$ en Présence de Symétrie. V, 174 pages. 1976.

Vol. 511: Séminaire de Probabilités X. Proceedings 1974/75. Edité par P. A. Meyer. VI, 593 pages. 1976.

Vol. 512: Spaces of Analytic Functions, Kristiansand, Norway 1975. Proceedings. Edited by O. B. Bekken, B. K. Øksendal, and A. Stray. VIII, 204 pages. 1976.

Vol. 513: R. B. Warfield, Jr. Nilpotent Groups. VIII, 115 pages. 1976.

Vol. 514: Séminaire Bourbaki vol. 1974/75. Exposés 453 – 470. IV, 276 pages. 1976.

Vol. 515: Bäcklund Transformations. Nashville, Tennessee 1974. Proceedings. Edited by R. M. Miura. VIII, 295 pages. 1976.

Vol. 516: M. L. Silverstein, Boundary Theory for Symmetric Markov Processes. XVI, 314 pages. 1976.

Vol. 517: S. Glasner, Proximal Flows. VIII, 153 pages. 1976.

Vol. 518: Séminaire de Théorie du Potentiel, Proceedings Paris 1972-1974. Edité par F. Hirsch et G. Mokobodzki. VI, 275 pages. 1976.

Vol. 519: J. Schmets, Espaces de Fonctions Continues. XII, 150 pages. 1976.

Vol. 520: R. H. Farrell, Techniques of Multivariate Calculation. X, 337 pages. 1976.

Vol. 521: G. Cherlin, Model Theoretic Algebra – Selected Topics. IV, 234 pages. 1976.

Vol. 522: C. O. Bloom and N. D. Kazarinoff, Short Wave Radiation Problems in Inhomogeneous Media: Asymptotic Solutions. V. 104 pages. 1976.

Vol. 523: S. A. Albeverio and R. J. Høegh-Krohn, Mathematical Theory of Feynman Path Integrals. IV, 139 pages. 1976.

Vol. 524: Séminaire Pierre Lelong (Analyse) Année 1974/75. Edité par P. Lelong. V, 222 pages. 1976.

Vol. 525: Structural Stability, the Theory of Catastrophes, and Applications in the Sciences. Proceedings 1975. Edited by P. Hilton. VI, 408 pages. 1976.

Vol. 526: Probability in Banach Spaces. Proceedings 1975. Edited by A. Beck. VI, 290 pages. 1976.

Vol. 527: M. Denker, Ch. Grillenberger, and K. Sigmund, Ergodic Theory on Compact Spaces. IV, 360 pages. 1976.

Vol. 528: J. E. Humphreys, Ordinary and Modular Representations of Chevalley Groups. III, 127 pages. 1976.

Vol. 529: J. Grandell, Doubly Stochastic Poisson Processes. X, 234 pages. 1976.

Vol. 530: S. S. Gelbart, Weil's Representation and the Spectrum of the Metaplectic Group. VII, 140 pages. 1976.

Vol. 531: Y.-C. Wong, The Topology of Uniform Convergence on Order-Bounded Sets. VI, 163 pages. 1976.

Vol. 532: Théorie Ergodique. Proceedings 1973/1974. Edité par J.-P. Conze and M. S. Keane. VIII, 227 pages. 1976.

Vol. 533: F. R. Cohen, T. J. Lada, and J. P. May, The Homology of Iterated Loop Spaces. IX, 490 pages. 1976.

Vol. 534: C. Preston, Random Fields. V, 200 pages. 1976.

Vol. 535: Singularités d'Applications Differentiables. Plans-sur-Bex. 1975. Edité par O. Burlet et F. Ronga. V, 253 pages. 1976.

Vol. 536: W. M. Schmidt, Equations over Finite Fields. An Elementary Approach. IX, 267 pages. 1976.

Vol. 537: Set Theory and Hierarchy Theory. Bierutowice, Poland 1975. A Memorial Tribute to Andrzej Mostowski. Edited by W. Marek, M. Srebrny and A. Zarach. XIII, 345 pages. 1976.

Vol. 538: G. Fischer, Complex Analytic Geometry. VII, 201 pages. 1976.

Vol. 539: A. Badrikian, J. F. C. Kingman et J. Kuelbs, Ecole d'Eté de Probabilités de Saint Flour V-1975. Edité par P.-L. Hennequin. IX, 314 pages. 1976.

Vol. 540: Categorical Topology, Proceedings 1975. Edited by E. Binz and H. Herrlich. XV, 719 pages. 1976.

Vol. 541: Measure Theory, Oberwolfach 1975. Proceedings. Edited by A. Bellow and D. Kölzow. XIV, 430 pages. 1976.

Vol. 542: D. A. Edwards and H. M. Hastings, Čech and Steenrod Homotopy Theories with Applications to Geometric Topology. VII, 296 pages. 1976.

Vol. 543: Nonlinear Operators and the Calculus of Variations, Bruxelles 1975. Edited by J. P. Gossez, E. J. Lami Dozo, J. Mawhin, and L. Waelbroeck, VII, 237 pages. 1976.

Vol. 544: Robert P. Langlands, On the Functional Equations Satisfied by Eisenstein Series. VII, 337 pages. 1976.

Vol. 545: Noncommutative Ring Theory. Kent State 1975. Edited by J. H. Cozzens and F. L. Sandomierski. V, 212 pages. 1976.

Vol. 546: K. Mahler, Lectures on Transcendental Numbers. Edited and Completed by B. Diviš and W. J. Le Veque. XXI, 254 pages. 1976.

Vol. 547: A. Mukherjea and N. A. Tserpes, Measures on Topological Semigroups: Convolution Products and Random Walks. V, 197 pages. 1976.

Vol. 548: D. A. Hejhal, The Selberg Trace Formula for PSL (2, IR). Volume I. VI, 516 pages. 1976.

Vol. 549: Brauer Groups, Evanston 1975. Proceedings. Edited by D. Zelinsky. V, 187 pages. 1976.

Vol. 550: Proceedings of the Third Japan – USSR Symposium on Probability Theory. Edited by G. Maruyama and J. V. Prokhorov. VI, 722 pages. 1976.